Nonlinear Optical Properties of Organic and Polymeric Materials

Nonlinear Optical Properties of Organic and Polymeric Materials

David J. Williams, EDITOR
Eastman Kodak Research Laboratories

Based on a symposium sponsored
by the ACS Division
of Polymer Chemistry
at the 184th Meeting
of the American Chemical Society,
Kansas City, Missouri,
September 12–17, 1982

ACS SYMPOSIUM SERIES **233**

AMERICAN CHEMICAL SOCIETY
WASHINGTON, D.C. 1983

Library of Congress Cataloging in Publication Data

Nonlinear optical properties of organic and polymeric materials.
(ACS symposium series, ISSN 0097-6156; 233)

"Based on a symposium sponsored by the ACS Division of Polymer Chemistry, at the 184th meeting of the American Chemical Society, Kansas City, Missouri, September 12-17, 1982."

Bibliography: p.
Includes index.

1. Protective coatings—Optical properties—Congresses.

I. Williams, David J., 1943– . II. American Chemical Society. Division of Polymer Chemistry. III. Series.

TP934.N66 1983 620.1'9204295 83-15514
ISBN 0-8412-0802-6

PRINTED IN THE UNITED STATES OF AMERICA

FOREWORD

The ACS SYMPOSIUM SERIES was founded in 1974 to provide a medium for publishing symposia quickly in book form. The format of the Series parallels that of the continuing ADVANCES IN CHEMISTRY SERIES except that in order to save time the papers are not typeset but are reproduced as they are submitted by the authors in camera-ready form. Papers are reviewed under the supervision of the Editors with the assistance of the Series Advisory Board and are selected to maintain the integrity of the symposia; however, verbatim reproductions of previously published papers are not accepted. Both reviews and reports of research are acceptable since symposia may embrace both types of presentation.

CONTENTS

PREFACE

DEVELOPMENTS IN THE FIELD of nonlinear optics hold promise for important applications in optical information processing, telecommunications, and integrated optics. Because of the emergence of this field from solid-state physics, in which inorganic semiconductors, insulators, and crystals constituted a major part of the scientific base, the early experimental and theoretical investigations were primarily concerned with materials from these classes. Particularly important is the more recent recognition that organic and polymeric materials with large delocalized π–electron systems may exhibit extremely large nonlinear responses, in many cases much larger than their inorganic counterparts. In addition, the properties of organic or polymeric materials may be varied to optimize adjunct properties, e.g., mechanical, thermal stability, and laser damage threshold, but preserve the electronic interactions responsible for the nonlinear optical effect.

Thin films of organic or polymeric materials with large second-order nonlinearities in combination with Si-based electronic circuitry offer the possibility of novel phenomena and devices for laser modulation and deflection, information control in optical circuitry, light valves, optical bistability, and all optical switches. Other novel processes occurring through the third-order nonlinearity such as degenerate four-wave mixing, whereby real-time processing of optical fields takes place, may find utility in such diverse fields as optical communications and integrated circuit fabrication. Of particular importance for conjugated organic systems is the fact that the origin of the nonlinear effects is the polarization of the π–electron cloud as opposed to displacement or rearrangement of nuclear coordinates found in inorganic materials. Thus the potential utility of these materials for very high frequency application contrasts with the bandwidth limitations of conventional inorganic electro-optic materials.

Although the field of nonlinear optics has traditionally been the stronghold of the physics and electrical engineering disciplines, if many of the potential applications are to be realized materials science and chemistry must play a role in its future development. A parallel can be drawn between the multidisciplinary effort that has been responsible for the tremendous progress in integrated electronic circuitry in recent years, and the need to combine the development of nonlinear media with sophisticated physical characterization and exploration of new nonlinear phenomena, in order for progress to be sustained.

This volume presents the most recent findings of researchers who have established the state of the art in preparation and characterization of new organic and polymeric nonlinear optical media, and reviews progress in this emerging combination of the fields of chemistry and physics. Because of the extremely large second and third nonlinear optical responses exhibited by conjugated organic systems and the unique ability of organic and polymer chemistry to combine electronic functionality with other useful properties, the symposium on which this volume is based focused on progress and problems in this segment of the field. An additional aim of this volume is to stimulate progress in the design of new materials and fabrication methods that could lead to broad technological utility of many important nonlinear optical phenomena.

To introduce the subject matter of this volume, the fundamental concepts of nonlinear optics and their relationship to chemical structures are briefly summarized. In the dipolar approximation, the polarization induced in an atom or molecule by an external field **E** can be written as

$$\mathbf{P} = \alpha \cdot \mathbf{E} + \beta \cdot \cdot \mathbf{EE} + \gamma \cdot \cdot \cdot \mathbf{EEE} + \cdots$$

where the vector quantities **P** and **E** are related by the tensor quantities α, β, and γ, which are often referred to as the polarizability, hyperpolarizability, and second hyperpolarizability, respectively. Similarly the polarization induced in macroscopic or bulk media can be expressed as

$$\mathbf{P} = \chi^{(1)} \cdot \mathbf{E} + \chi^{(2)} \cdot \cdot \mathbf{EE} + \chi^{(3)} \cdot \cdot \cdot \mathbf{EEE} + \cdots$$

where the coefficients $\chi^{(1)}$, $\chi^{(2)}$, and $\chi^{(3)}$ are similar in meaning to their microscopic counterparts. In this formalism the even order tensors β and $\chi^{(2)}$ are zero in centrosymmetric media, whereas the odd order tensors do not have symmetry restrictions associated with them. The molecular quantities α and γ are related to their macroscopic counterparts through summations over the number of contributing atoms or molecules per unit volume and are corrected for local contributions from neighboring molecular fields. Conversely, a molecule with an asymmetric charge distribution and therefore a nonzero β may exist in a centrosymmetric crystal or orientationally averaged molecular environment such as a liquid or amorphous polymer. Therefore, it may exhibit a vanishingly small value of $\chi^{(2)}$. To determine $\chi^{(2)}$ experimentally, the detailed nature of the propagation of the incident field and induced fields in the medium of interest must be carefully considered. For instance, conditions may exist where orientational averaging may not lead to zero $\chi^{(2)}$ (e.g., when the size of a noncentrosymmetric domain in an ensemble of orientationally averaged domains approaches or exceeds the wavelength of light). (*See* Chapters 2 and 4).

The nonlinear coefficients are determined by the detailed nature of the electronic environment of the medium, its symmetry for even order coefficients, and the exact nature of the interacting field components. For instance, molecules and molecular solids with delocalized π systems exhibit much larger coefficients than covalently bonded molecules and solids. The ability to tailor charge asymmetry, control the enhancement of the nonlinear polarization from electronic dispersion, and provide optical transparency in regions of the spectrum compatible with useful light sources are convenient tools of the physical organic chemist. These factors represent enormous potential for designing materials with specific properties for certain applications. For example, second harmonic generation, which is often designated by the nonlinear coefficient $\chi^{(2)}$ $(-2\omega,\omega,\omega)$, can be extremely efficient in crystals of 2-methyl-4-nitroaniline. (The argument of $\chi^{(2)}$ indicates that the two interacting fields ω produce polarization at 2ω, and the negative sign indicates that the wave vectors of the interacting fields must sum to zero.) The molecule has a strong intramolecular charge-transfer optical transition in the vicinity of 430 nm and crystallizes in a polar space group. Therefore, this material is useful for converting 1.06-μm radiation from a Nd^{3+} YAG laser to 530 nm. Similarly, linear electro-optic (or Pockets effect) is represented as $\chi^{(2)}$ $(-\omega,\text{o},\omega)$, which indicates interaction of the optical and DC field components. In organic systems these coefficients are very similar in magnitude because the polarization of the π system is very insensitive to frequencies below that for the first optical transitions.

The situation for $\chi^{(3)}$ becomes complicated very quickly because the odd order coefficients can produce responses at the input frequencies as well as resonant effects like two-proton absorption and various Raman active resonances. At least 20 different processes occurring through $\chi^{(3)}$ have been observed. The extremely large $\chi^{(3)}$ responses found in conjugated organic systems and polymers combined with design flexibility of organic and polymeric materials may lead to important applications in optical information and signal processing.

With the burgeoning of interest in this field, the publication of this book is very timely. An effort was made to include contributions from all the major research efforts in the world concerned with the development of organic and polymeric materials for nonlinear optical applications. The chapters in this book serve as a comprehensive view of the state of the art in this field, yet sufficient introductory and review material is included to render it useful for researchers interested in entering this field.

DAVID J. WILLIAMS
Eastman Kodak Research Center
Rochester, NY

July 1983

1

Molecular Optics: Nonlinear Optical Properties of Organic and Polymeric Crystals

A. F. GARITO, K. D. SINGER[1], and C. C. TENG

University of Pennsylvania, Department of Physics and Laboratory for Research on the Structure of Matter, Philadelphia, PA 19104

Studies of the basic physics of nonlinear second order optical processes in solids remain critically important to the field of nonlinear optics and to further advances in telecommunications technologies. This fact receives continued emphasis today in problems of current interest such as phase conjugation and optical bistability. Research activities have centered on second order optical processes occurring in inorganic dielectric insulators and semiconductors primarily because of the important scientific achievements and resultant large body of available information from earlier studies of piezoelectric and ferroelectric properties and semiconductor transport. Consequently, nearly all of the nonlinear optical materials widely under study are inorganic solids. Recently, however, much interest has focused on organic and polymeric solids basically due to their exceptionally large nonlinear second order optical properties and the promise of virtually unlimited numbers of crystalline structures.(1-9)

This article by its nature is limited to a brief summary of recent developments in the authors' research on fundamental studies of nonlinear second order optical properties of organic and polymeric crystals and films. In particular, we describe several important results from these studies that have systematically led to microscopic understanding of the nature of the electronic states and origin of the large nonlinear optical responses in organic solids and crystalline polymers while reporting recent progress in on-going studies of polymers.

[1] Current address: Western Electric Laboratories, Princeton, NJ 08540.

0097-6156/83/0233-0001$07.50/0

Second Harmonic Generation and Linear Electrooptic Effect in Solids

Nonlinear second order optical properties such as second harmonic generation and the linear electrooptic effect arise from the first non-linear term in the constitutive relation for the polarization P(t) of a medium in an applied electric field E(t) = E cos ωt.

$$P(t) = \varepsilon_0 \chi^{(1)} E(t) + \varepsilon_0 \chi^{(2)} E^2(t) + \varepsilon_0 \chi^{(3)} E^3(t) + \ldots \quad (1)$$

The first and third order terms in odd powers of the applied electric field are present for all materials. In the second order term, a polarization is induced proportional to the square of the applied electric field, and the nonlinear second order optical susceptibility $\chi^{(2)}(-\omega_3;\ \omega_1,\ \omega_2)$ must, therefore, vanish in crystals that possess a center of symmetry. In addition to the noncentrosymmetric structure, efficient second harmonic generation requires crystals to possess propagation directions where the crystal birefringence cancels the natural dispersion leading to phase matching.

For crystalline solids, comparative quantities for second harmonic generation and the linear electrooptic coefficient are given by Miller's delta and the polarization optic coefficient f. The quantity δ is defined by the relation

$$d_{ijk} = \varepsilon_0 \chi^{(1)}_{ii} \chi^{(1)}_{jj} \chi^{(1)}_{kk}\ \delta_{ijk} \quad (2)$$

where terms such as $\chi^{(1)}_{i}$ are the linear susceptibility components, and d_{ijk}, the second harmonic coefficient, is defined through

$$\chi^{(2)}_{ijk}(-2\omega;\ \omega, \omega) = 2\ d_{ijk}(-2\omega;\ \omega,\ \omega) \quad (3)$$

The polarization optic coefficient f is defined by the expression

$$r_{ijk} = \varepsilon_0 (\varepsilon_k - 1) f_{ijk} \quad (4)$$

where ε_k is the dielectric constant, and r_{ijk}, the linear electrooptic coefficient, is defined as

$$\chi^{(2)}_{ijk}(-\omega;\ \omega,\ 0) = 1/2\ n_{ii}^2 n_{jj}^2 r_{ijk}(-\omega;\ \omega,\ 0) \quad (5)$$

where n_{ii} is the index of refraction.

Figure 1 shows the standard quantities for selected organic and polymeric examples whose δ and f

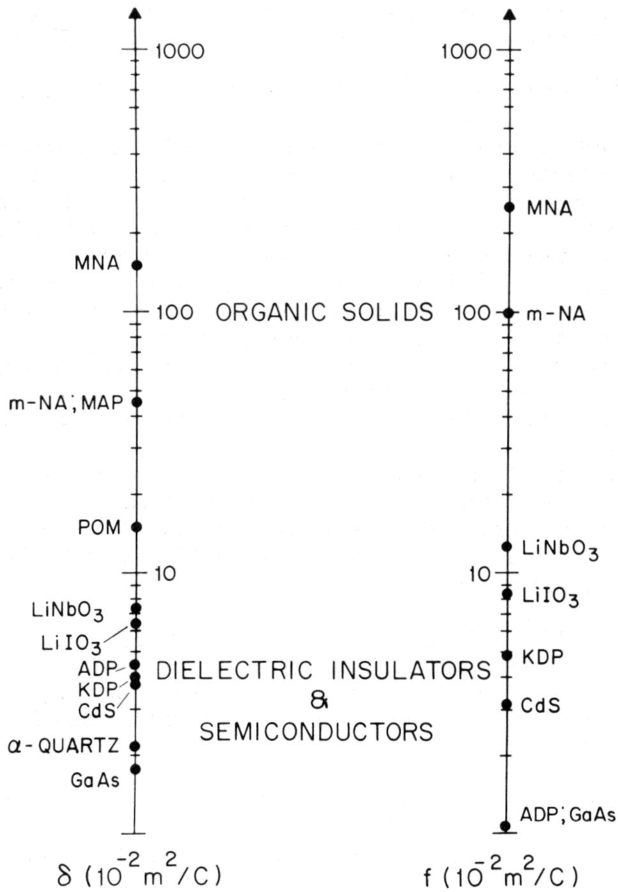

Figure 1. Comparative quantities for selected tensor components of second harmonic generation (left) and the linear electro-optic effect (right) (measured at 1.06 μm wavelength). (The strain free quantity, f, *was measured at 0.633 μm wavelength except in the case of GaAs which was measured at = 0.9 μm). (Reproduced with permission from Ref. 8. Copyright 1982, Laser Focus.)*

susceptibility $\chi^{(2)}_{111}$ $(-\omega;\ \omega,\ 0)$ of 540 $\pm$ 100 x 10^{-12} m/V. The finding that the electronic contribution dominates the frequency range from zero to optical frequencies has also been reported in other organic molecular solids and appears to be a property common to optically nonlinear organic solids. This is an outstanding property in other respects in that such electronic excitations provide extremely fast intrinsic switching times of order 10^{-14} seconds.

The π-electron excitations are viewed as occuring on molecular sites weakly coupled to their neigbors and providing sources of nonlinear optical response through the on-site microscopic second order nonlinear electronic susceptibility β_{ijk} $(-\omega_3;\omega_1,\omega_2)$. In second order perturbation theory with the perturbing Hamiltonian $H' = e\vec{E}^{\omega}\cdot\vec{r}\cos\omega t$, and both the fundamental and created combined frequencies below electronic resonances but well above vibrational and rotational modes, can be expressed as

$$\beta_{ijk}+\beta_{ikj} =$$

$$-\frac{e^3}{4\hbar^2}\Bigg[\sum_{\substack{n\neq n'\\ n\neq g\\ n'\neq g}}(r_{gn'}{}^{j}r_{n'n}{}^{i}r_{gn}{}^{k}+r_{gn'}{}^{k}r_{n'n}{}^{i}r_{gn}{}^{j})\left(\frac{1}{(\omega_{n'g}-\omega)(\omega_{ng}+\omega)}+\frac{1}{(\omega_{n'g}+\omega)(\omega_{ng}-\omega)}\right)$$

$$+(r_{gn'}{}^{i}r_{n'n}{}^{j}r_{gn}{}^{k}+r_{gn'}{}^{i}r_{n'n}{}^{k}r_{gn}{}^{j})\left(\frac{1}{(\omega_{n'g}+2\omega)(\omega_{ng}+\omega)}+\frac{1}{(\omega_{n'g}-2\omega)(\omega_{ng}-\omega)}\right) \qquad (6)$$

$$+(r_{gn'}{}^{j}r_{n'n}{}^{k}r_{gn}{}^{i}+r_{gn'}{}^{k}r_{n'n}{}^{j}r_{gn}{}^{i})\left(\frac{1}{(\omega_{n'g}-\omega)(\omega_{ng}-2\omega)}+\frac{1}{(\omega_{n'g}+\omega)(\omega_{ng}+2\omega)}\right)$$

$$+4\sum_{n}[r_{gn}{}^{j}r_{gn}{}^{k}\Delta r_{n}{}^{i}(\omega_{ng}{}^{2}-4\omega^2)+r_{gn}{}^{i}(r_{gn}{}^{k}\Delta r_{n}{}^{j}+r_{gn}{}^{j}\Delta r_{n}{}^{k})(\omega_{ng}{}^{2}+2\omega^2)]$$

$$\times\frac{1}{(\omega_{ng}{}^{2}-\omega^2)(\omega_{ng}{}^{2}-4\omega^2)}\Bigg]$$

where summations are over complete sets of eigenstates $\langle n|$ and $\langle n'|$ of the unperturbed system, $r^{i}_{nn} = \langle n|r^{i}|n'\rangle$, $\Delta r^{i}_{n} = r^{i}_{nn} - r^{i}_{gg}$, and $h\omega_{ng}$ the difference between excited and ground state energies.

Thus, since intramolecular bonding interactions in the solid are much stronger than relatively weak intermolecular van der Waals interactions, each molecular unit is essentially an independent source of nonlinear response, arrayed in an acentric cystal structure, and coupled to its neighbors mainly through weak local fields. In the rigid lattice gas approximation, the macroscopic susceptibility $\chi^{(2)}$ is expressed as

$$\chi^{(2)}_{ijk}(-\omega_3;\omega_1,\omega_2) = Nf^{\omega_3}f^{\omega_2}f^{\omega_1}\langle\beta_{ijk}(-\omega_3;\omega_1,\omega_2)\rangle \qquad (7)$$

values are orders of magnitude larger than those of commmon inorganic dielectric insulators and semiconductors(9). The δ and f values for the inorganic materials lie within a factor of two of their respective mean values. In contrast, the organic solid MNA, for example, possesses δ and f values 50 times larger than those of KDP.

In each case, the major polarizable medium in the organic molecular solids and conjugated polymers is the π-electron system, and the origin of the unusally large values of δ and f, or equivalently $\chi^{(2)}$, lies in the excited π-electron states, particularly those possessing large charge correlations as will be described below. In general, π-electron systems exhibit optical excitations in the near ultraviolet, and optical transparency from the visible to the near infrared regions, leading to indices of refraction in the range 1.6 - 2.

Origin of $\chi^{(2)}$ and the Second Order Nonlinear Electronic Susceptibility β

Organic molecular units and conjugated polymer chains possessing π-electron systems usually form as centrosymmetric structures and thus, in the electric dipole approximation, would not show any nonlinear second order optical properties. The necessary acentric structure may be provided by first distorting the π-electron system by interaction with strong electron donor and acceptor groups. The case of MNA, whose x-ray determined molecular structure is shown in Figure 2, serves as an important example(1,2). In this system, the benzene ring is substituted by amine (NH_2) and methyl (CH_3) donor groups and oppositely by a nitro (NO_2) acceptor group. The resultant molecular ground state dipole moment (6.94 D) points nearly axially across the ring from the NH_2 to the NO_2 group. MNA crystallizes in the solid state in a highly acentric monoclinic (C_c) structure. X-ray studies reveal a unique projection (Figure 3) along the crystal polar axis where individual microscopic dipole moments of the molecular units are all aligned along the polar axis. It is this single polar orientation that is principally responsible for the exceptionally large δ and f values of MNA(2).

The dominance of the π-electron excitations in MNA is demonstrated by the experimental finding that along the polar axis the second harmonic susceptibility $\chi^{(2)}_{111}(-2\omega;\omega,\omega)$ of $500 \pm 125 \times 10^{-12}$ m/V is the same as the linear electrooptic

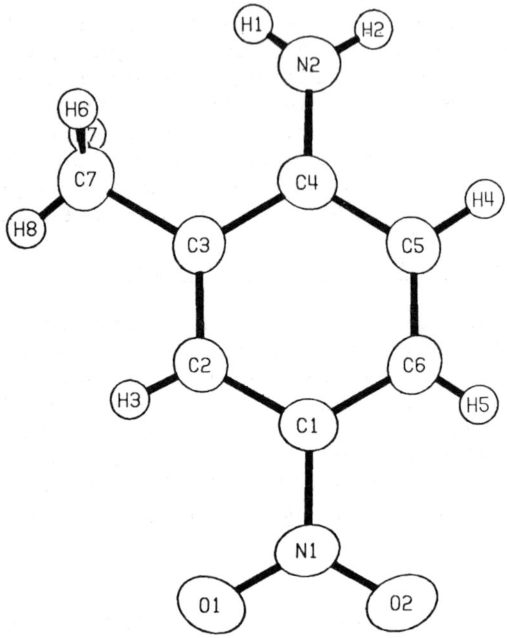

Figure 2. The molecular structure of 2-methyl-4-nitroaniline (MNA) as determined by x-ray crystallography (2). *Key: C, carbon; N, nitrogen; O, oxygen; and H, hydrogen.*

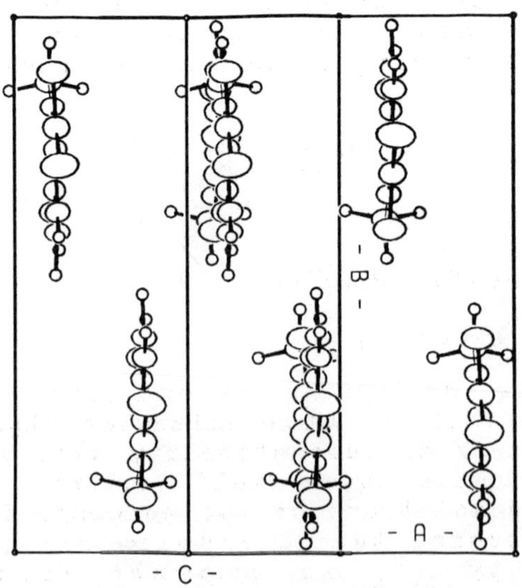

Figure 3. Projection showing the independent molecular units of the crystallographic unit cell of MNA down the polar axis (2).

where N is the number of sites per unit volume and $f^{\omega i}$ represents local field corrections. Thus, unlike inorganic semiconductors and dielectric insulators, the problem of understanding the origin of the large macroscopic $\chi^{(2)}$ is reduced to experimental and theoretical studies of the corresponding microscopic β of single molecular units, or polymer chains, comprising the optically nonlinear organic solid. Such studies are illustrated in the next sections.

DC Induced Second Harmonic Generation

The value of β and its dispersion for a molecule, or polymer chain, can be experimentally determined by DC induced second harmonic generation (DCSHG) measurements of liquid solutions(9-12). The experimental arrangement requiring an applied DC field E^o to remove the natural center of inversion symmetry of the solution is described in Figure 4. The second harmonic polarization of the solution is expressed as

$$P_i^2 = \Gamma_{ijkl} E_j^{\omega} E_k^{\omega} E_l^o \qquad (8)$$

where the product $\Gamma_{ijkl} E_l^o$ of the third order nonlinear optical susceptiblity of the solution and the DC field E_l^o acts as an effective second harmonic susceptibility $\chi_{ijk}^{(2)}(-2\omega;\omega,\omega)$. Γ_{ijkl} contains β_{ijk} and the smaller microscopic third order susceptibility γ_{ijkl} for electronic excitations and the hyper Raman effect. After statistical averaging over the liquid and with all electric fields aligned parallel, Γ_{ijkl} is given by

$$\Gamma_{1111} = Nf^o(f^{\omega})^2 f^{2\omega}(\gamma - \frac{\beta\mu}{5k_BT}) \qquad (9)$$

and

$$\overline{\beta\mu} \equiv 1/3 \sum_{u,v}(2\beta_{uuv}\mu_v + \beta_{uvv}\mu_u)$$

$$\gamma \equiv 1/15 \sum_{u,v}(2\gamma_{uuvv} + \gamma_{uvvu})$$

where β is the molecular second order nonlinear optical susceptibility, μ the molecular dipole moment; and γ the third order susceptibility with $\gamma << \overline{\beta\mu}/k_BT$, and f are quantities local field factors.

For dilute solutions of the molecule dissolved in a suitable solvent, the DC induced second harmonic susceptibility Γ_L is a combination of the susceptibilities Γ_o and Γ_1 from the solvent and solute, respectively,

$$\Gamma_L = \Gamma_o + \Gamma_1$$

$$= (N_o F_o \beta_{xo} \mu_o + N_1 F_1 \beta_{x1} \mu_1)\ 1/5k_B T \qquad (10)$$

where N_i is the number density and the local field factors $F_i = f_i^o f_i^{2\omega} (f_i^{\omega})^2$. To minimize solute-solute interactions, β_{x1} is obtained in the infinite dilution limit using Onsager local field corrections(9)

$$\frac{(2\varepsilon_o + n_1^2)(2n_o^2 + n_1^2)^3 M_1}{(n_1^2 + 2)^4 n_o^6 \varepsilon_o} \left\{ \left(v_o \frac{\partial \Gamma_L}{\partial w} \right)_o + \Gamma_o \left(\frac{\partial v}{\partial w} \right)_o + \right.$$

$$\left. v_o\Gamma_o - v_o\Gamma_o \left(\left(\frac{1}{n_o^2} \frac{\partial n^2}{\partial w} \right)_o + \left(\frac{1}{\varepsilon_o} - \frac{2}{2\varepsilon_o + n_o^2} \right) \frac{\partial \varepsilon}{\partial w} \right)_o \right\} =$$

$$N_A \frac{\beta_{x1}\mu_1}{5kT} \qquad (11)$$

where ε_i is the static dielectric constant, v_i specific volume, n_i refractive index, M molecular weight and w weight fraction. Data for p-nitroaniline (PNA) dissolved in dioxane are given in Figure 5 to illustrate extrapolation of Γ_L values to the infinite dilution limit.

The intrinsic value for β_{x1} is obtained by then accounting for dipole-dipole interactions between the solute and solvent. The dominant effect is a shift in the excitation energy $\hbar\omega_{ng}$ of the solute molecule caused by changes in the ground and excited state dipole moments of the solute molecule

$$\Delta\omega_{ng} = \frac{f(n)}{\hbar a^3} (\mu_n^2 - \mu_g^2) \qquad (12)$$

where a is the cavity radius containing the solvent molecule, f(n) a standard function of the solvent refractive index, and $\mu_i = -e\, r_{ii}^x$. As shown in equation 6, the excitation energy $\hbar\omega_{ng}$ is an important factor determining the magnitude of β and correspondingly its dispersion, especially for near resonant responses. The fundamental case PNA is discussed as an illustration in the next section.

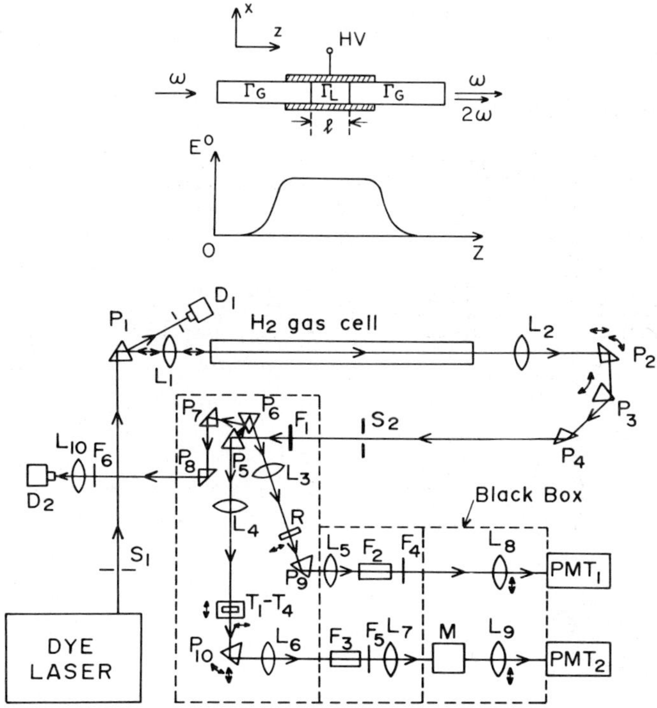

Figure 4. The sample cell arrangement in the DCSHG experiment, where the sample solution was inserted between two glass slips (top), and the optical design for the DCSHG dispersion experiment, where the compressed H_2 gas medium was pumped by a tunable pulsed dye laser source for Stokes generation by stimulated Raman scattering (bottom). (E° is the static electric field.) Key: P_1–P_{10}, beam guiding prisms; P_3, Stokes selecting prism; L_1–L_{10}, focusing lens; F_1–F_5, optical filters; M, monochromator; T_1–T_4, translational stages on which the sample cell is positioned; R, reference quartz crystal; and PMT_1 and PMT_2, broad band photomultiplier tubes. (Reproduced with permission from Ref. 12. Copyright 1983, Phys. Rev. Lett.*)*

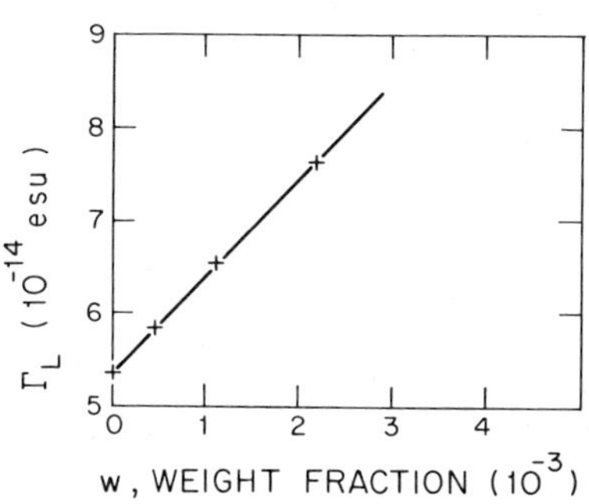

Figure 5. Typical data points of Γ_L vs. w *in the extrapolation procedure for PNA dissolved in dioxane (λ = 1.06 μm). (Reproduced with permission from Ref. 12. Copyright 1983,* Phys. Rev. Lett.*)*

Theoretical Calculations of β and Charge Correlated π-Electron States

The value of β and its dispersion can be theoretically calculated from equation 6, provided a complete set of electron states of the system is known. Such quantum mechanical calculations have been developed based on molecular Hartree-Fock theory including configuration interactions(13). A detailed theoretical analysis of β and contributing π-electron states has been presented for several important molecular structures.

The fundamentally important case p-nitroaniline (PNA), the high symmetry parent molecular unit of MNA, presents an excellent example for both experimental methodology and theoretical analysis. In the C_{2v} symmetry for PNA and second harmonic generation, the measured principal component β_x is given by

$$\beta_x = \beta_{xxx} + 1/3[\beta_{xyy} + \beta_{xzz} + 2\beta_{yyx} + 2\beta_{zzx}]$$

where the x direction is along the molecular dipole axis.

The self-consistent field calculations(13) are carefully guided by experiment through the following three step procedure: (1) determination of the many-body electronic ground state of the molecular system as an antisymmetrized product of the one-electron eigenfunction solutions of the Hartree-Fock equation using an all valence electron semiempirical parameterization. The solutions are directly compared to gas phase experimental photoemission results; (2) accounting for correlations by a configuration interaction (CI) calculations to form the lowest energy excited states and transition moments of the molecule. These results are verified by comparison of calculated energies and oscillator strengths with gas phase singlet-singlet excitation spectra; and (3) calculation of the tensor components of the molecular second-order optical susceptibility using perturbation theory.

The configuration interaction calculations using the PNA singlet excited states (4.2, 4.37, 4.38, 5.57, 6.17, 6.63, 6.80, 7.06, 7.49 eV) showed that the measured β value is dominated by virtual excitations to a highly charge correlated π-electron excited state at $\hbar\,\omega_{ng}$ of 4.37 eV corresponding to the first major transition in the singlet-singlet excitation spectrum of PNA. This many body excited state is composed mainly of an A_1 configuration in which an electron is promoted from the highest energy single electron orbital occupied in the ground state to the lowest energy single electron orbital that is unoccupied in the ground state. The corresponding electron density contour diagrams are shown in Figure 6(13).

The many body excited state clearly demonstrates the transfer of electron density from the region of the NH_2 donor group across the ring to the NO_2 acceptor group and its neighboring sites, giving the excited state a highly charge correlated nature along the polar X axis of the molecule. The change in dipole moment of the excited state with respect to the ground state $-e\Delta r_n^x$ was calculated to be 5.8D, and the transition dipole moment $-e\Delta r_{gn}^x$ 5.5D. Virtual excitations to this state account for over 90% of the magnitude of β_x of PNA as well as determine its positive sign.

The frequency dependent theoretical gas phase value β_x^g and the experimental values β_x^{exp} in the infinite dilution limit for PNA and MNA in dioxane are listed in Table 1 for several experimental fundamental frequencies(12). The dispersion of β_x^{exp} for PNA in dioxane is plotted in Figure 7. As the fundamental frequency is increased, β_x^{exp} for PNA rapidly increases smoothly as 2ω approaches the excitation frequency ($\omega_{ng} \sim 3.5$ eV/$\hbar$) which is the first major optical absorption peak of PNA in dioxane corresponding to excitations to the important second excited state. The data for MNA show the same behavior.

The difference between the experimental β_x^{exp} values and calculated gas phase β_x^g results for PNA (Table I) is due to the solute-solvent interactions that result in solvent induced shifts of the singlet-singlet excitation energies readily observable in solution absorption spectra(12). As expressed in equation 12, the shift in the excitation energy ($\hbar\omega_{ng}$) is caused by changes in the difference between the PNA ground and excited state dipole moments which in turn is an important quantity determining the magnitude of β_x.

The measured shifts in the singlet-singlet excitation spectrum of PNA in dioxane provide a new set of state energies. These energies were used in equation 6 to yield the calculated dispersion curve (solid line) shown in Figure 7(12). Within experimental error, the agreement between experiment and theory is quite satisfactory. The results for the

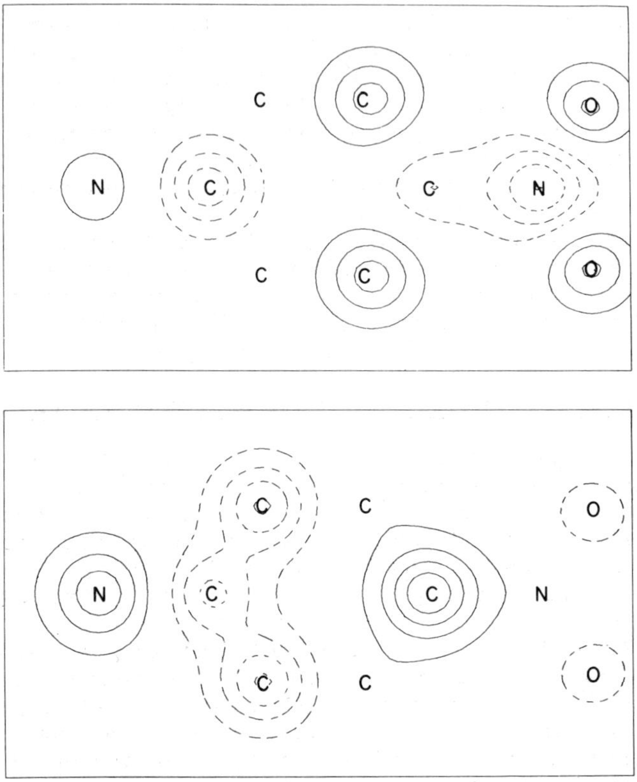

Figure 6. The contour diagrams of the highest occupied (bottom) and the lowest unoccupied (top) molecular orbitals for PNA. (Reproduced with permission from Ref. 13. Copyright 1979, Phys. Rev.*)*

Table I. Experimental and Gas Phase Theoretical β_x values of PNA and MNA.

λ (μm)	$\hbar\omega$(eV)	β_x^{exp}(10^{-30}esu) PNA	β_x^{exp}(10^{-3-}esu) MNA	β_x^{g} (10^{-30}esu) PNA
1.907	0.650	9.6 ± 0.5	9.5 ± 0.5	5.7
1.370	0.905	11.8 ± 0.3	12.8 ± 0.5	6.4
1.060	1.170	16.9 ± 0.4	16.7 ± 0.5	7.7
0.909	1.364	25 ± 1.0	27 ± 1	9.2
0.830	1.494	40 ± 3.0	45 ± 4	10.7

β_x^{exp}: Experimental value in dioxane.

β_x^{g} : Gas phase calculated value from Ref. 13.

lower symmetry for MNA are virtually the same. This agreement demonstrates that the theoretical microscopic mechanism is essentially correct in describing the origin of the large second order nonlinear optical response of organic structures in terms of highly charge correlated π-electron states. There are no analog states in inorganic semiconductors and dielectric insulators.

In reference to Figure 3 for MNA crystals, the polar axes of the individual molecular sites are aligned with one another along the crystal polar axis. The microscopic components β_x add resulting in the large macroscopic $\chi^{(2)}_{111}$ following equation 7.

Disubstituted Diacetylene Single Crystal Polymers

Optically nonlinear single crystal polymers possessing three dimensional long range order can be formed by irreversible topotatic or lattice controlled polymerization of disubstituted diacetylene monomers in the solid state(8,14). An illustration of this process is given in Figure 8. As shown, disubstituted diacetylene monomer units crystallize as linear arrays, where R and R' represent optically nonlinear, acentric organic substituents. The polymerization reaction can be initiated by thermal annealing, uv, x-ray or electron beam irradiation, and in some cases, even mechanically, forming three dimensionally long range ordered crystalline polymer chains in a *trans*-planar configuration.

An important example of the solid state polymerization of disubstituted diacetylenes is 1,6-bis(2,4-dinitro-phenoxy)-2,4-hexadiyne (DNP) shown in Figure 9(15). DNP is unique among known diacetylenes in that it polymerizes in the solid state by thermal annealing but not under ordinary x-ray conditions, thus allowing for the first time direct x-ray determinations of the diacetylene monomer (Figure 9) as well as polymer (Figure 10) structures. As shown in Figure 11, the thermal polymerization of DNP single crystals exhibits an induction period during which polymerization is slow, followed by an autocatalytic polymerization regime in which apparent first order kinetics are obeyed(15). The measured heat of polymerization $-\Delta H_p$ is 134.7 $\pm$ 6 kJ/mole (32.5 $\pm$ 1.5 kcal/mole) demonstrating that the preferred thermodynamic ground state in the solid is indeed the polymer structure.

DNP provided the first successful example of the systematic design and synthesis of highly nonlinear

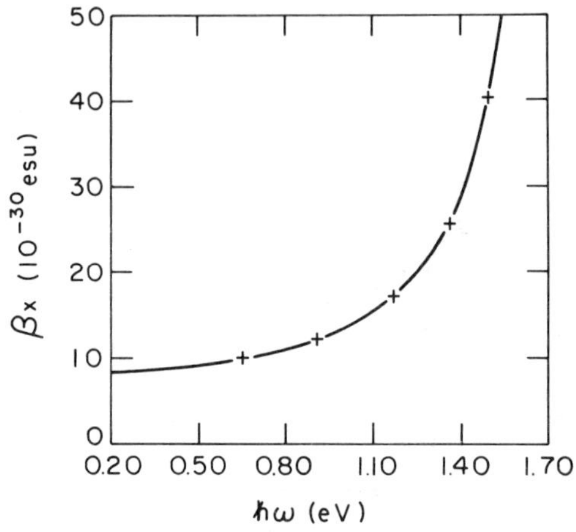

Figure 7. Frequency dependence of β_x for PNA. Key: +, experimental β_x^{exp} data points; and —, theoretical β_x^d curve accounting for solvent shift effect. Measured and calculated values agree even better than in Reference 12 due to the further refinements described in Reference 12.

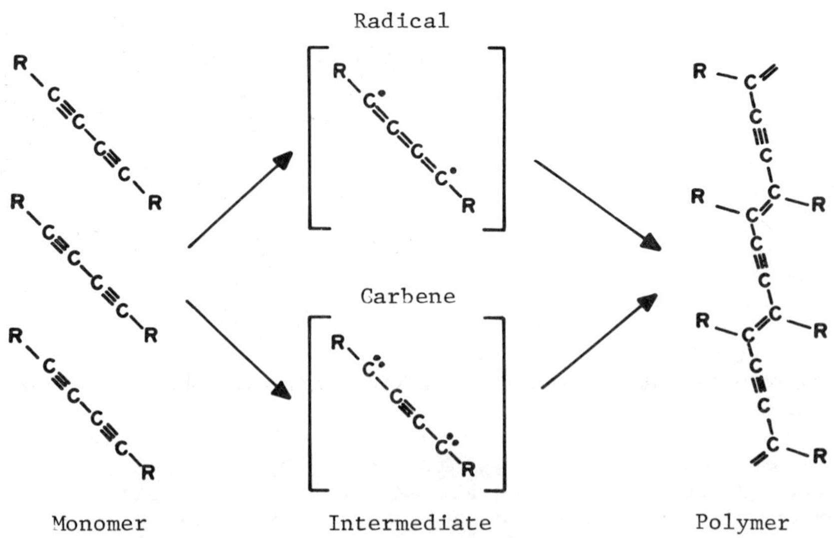

Figure 8. Solid-state polymerization of diacetylenes. A crystalline array of monomer units polymerizes through intermediate states to the final crystalline polymer chain.

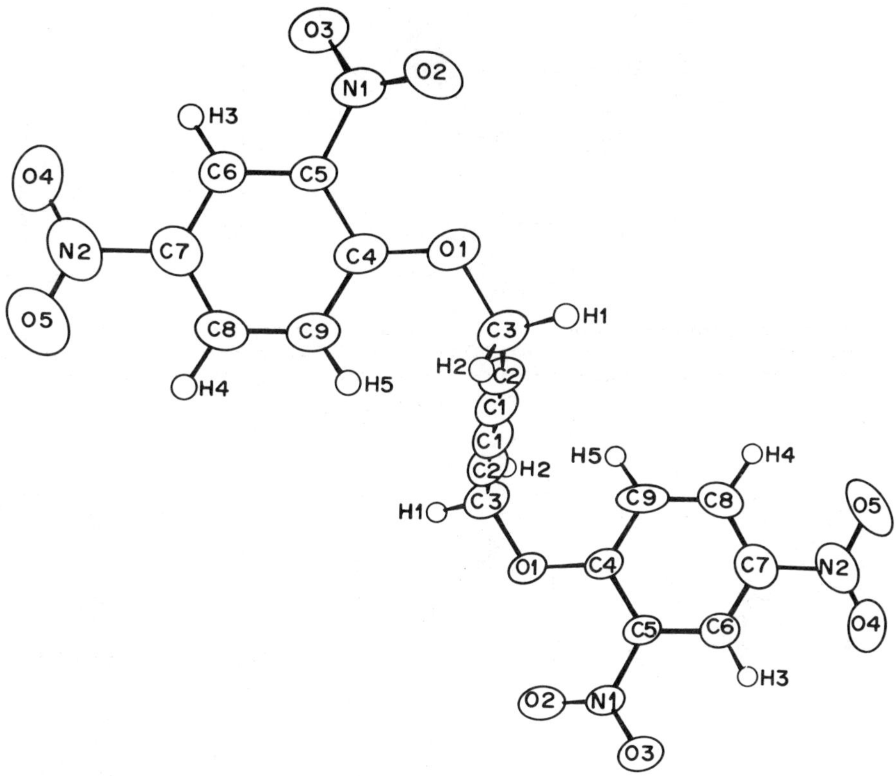

Figure 9. Molecular structure of 1,6-bis(2,4-dinitrophenoxy)-2,4-hexadiyne (DNP) (15).

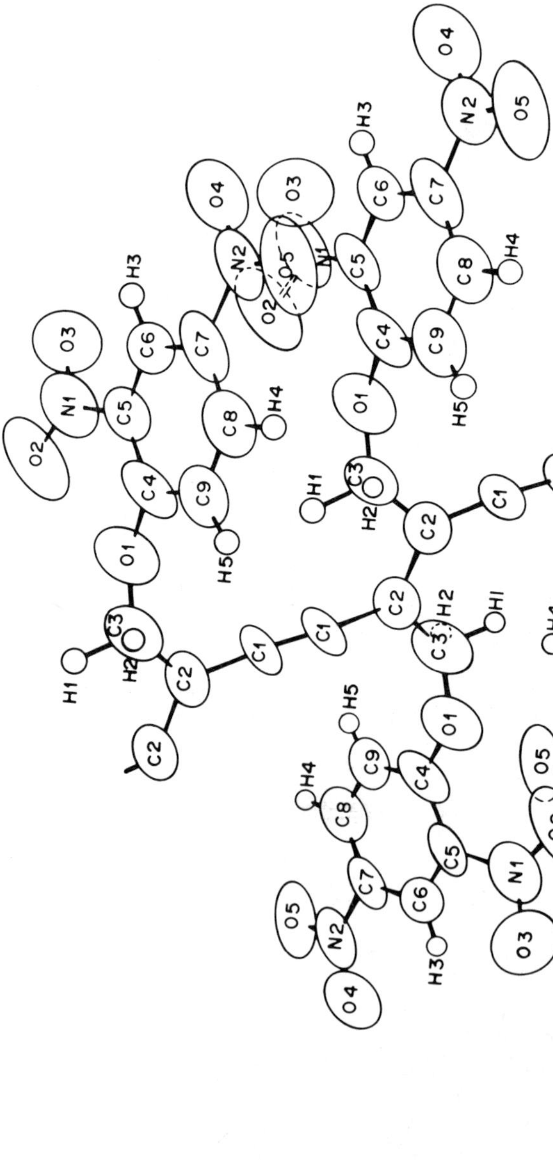

Figure 10. The polymer chain structure of 1,6-bis(2,4-dinitrophenoxy)-2,4-hexadiyne (DNP). Two repeating units are shown (15).

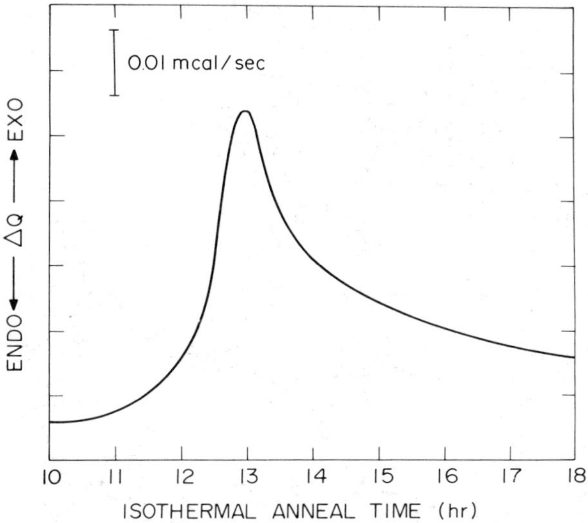

Figure 11. Spontaneous heat evolved vs. time of isothermal polymerization of DNP. (Reproduced with permission from Ref. 15. Copyright 1981, Makromol. Chem.*)*

molecular units placed in a strong polymer structure. Although centrosymmetric at room temperature, DNP crystals undergo an apparent ferroelectric transition at 30K (Figure 12) to a low temperature noncentrosymmetric structure(16) which provided important structural information for ultimately stabilizing other noncentrosymmetric diacetylene polymer structures at room temperature(7).

In addition to the π-electron system contained in the R and R' side groups, the polymer conjugated chain possesses a π-electron band structure that in general exhibits optical transparency from approximately the mid-visible region into the near infrared, leading to indices of refraction in the range 1.6-2. The natural structural anisotropy of the ordered polymer chains leads to large optical birefringence important to phase matching. Under polarized visible light, single crystals exhibit a lustrous metal-like reflectance for polarization along the chain axis direction and a rich highly colored reflectance for polarization perpendicular to the chain axis. Diacetylene polymer crystals also possess exceptional radiation and mechanical damage resistance; for instance, radiation damage thresholds have been observed as high as 1 GW/cm^2 for 25 nanosecond pulses at 1.89μ wavelength. Values of the Young's modulus and tensile strength along the chain are comparable to those of diamond.

Phase matched second harmonic generation in single crystal noncentrosymmetric polymers was first observed in MNA substituted diacetylene polymers(8). Two asymmetric monomer examples forming this new polymer class, NTDA and MNADA, are disubstiuted with MNA and an ethylamide group (Figure 13), and single crystals undergo solid state polymerization by thermal annealing, uv, and x-ray irradiation. The β values for the MNADA and NTDA monomers as measured by DCSHG are essentially the same as that of MNA alone. The result has been explained by localization of the π-electron contributions to β of the individual MNA, diacetylene, and ethylamide groups forming the monomer structure. The linking $(-CH_2-)_n$ methylene carbons act as π-electron blocking centers between the groups, and the observed β essentially results from the dominant MNA group.

In the solid state, the phase matched second harmonic signal increases with increased polymerization(7). Figure 14 shows preliminary data for the second harmonic intensity of NTDA microcrystals polymerized under x-ray radiation. As the polymer forms, the second harmonic intensity

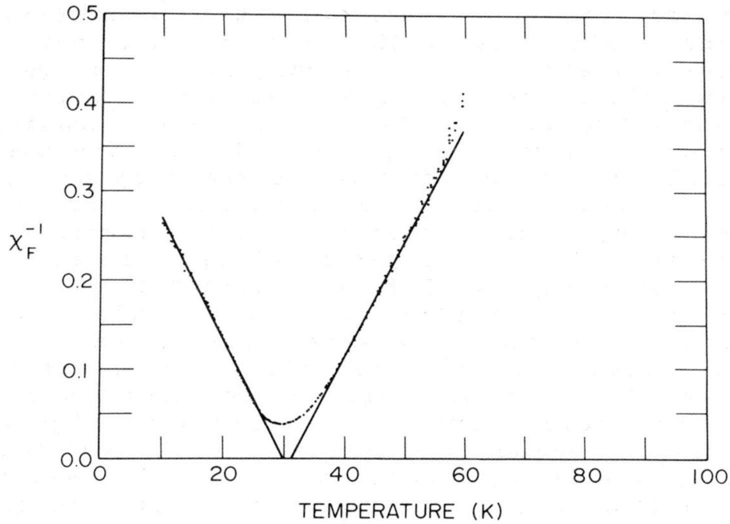

Figure 12. Temperature dependence of the inverse dielectric susceptibility (χ_F^{-1}) of DNP along the principal axis for polymerization. (Reproduced with permission from Ref. 16. Copyright 1980, Ferroelectric.*)*

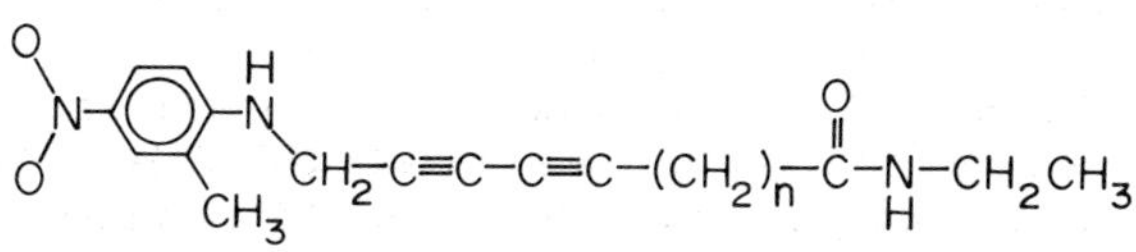

*Figure 13. Diacetylene monomer structures of MNADA (*n = *2) and NTDA (*n = *8).*

steadily increases to 10-15 times the $LiIO_3$ reference signal. The increase in signal intensity with increased polymer concentration can result from direct contributions to $\chi^{(2)}$ from the polymer π-electron band structure, or from more efficient phase matching between the fundamental and second harmonic waves as the anisotropic index of refraction changes. These aspects together with newly developed diacetylene structures exhibiting even higher second harmonic intensities remain at the focus of on-going studies.

Diacetylene Films and Optical Guided Wave Structures

In addition to studies of diacetylene single crystals, current research activities are focused on studies of the second $\chi^{(2)}$ and third $\chi^{(3)}$ order nonlinear optical responses of disubstituted diacetylene polymer films as active optical guided wave structures. Diacetylene polymers possess $\chi^{(3)}$ values comparable to germanium(17). In the first stage, three major questions are being addressed:
(i) use of the naturally large index of refraction of diacetylene polymers to satisfy the principle waveguiding requirement for the refractive index of the guide n_2 to be greater than that of the substrate n_1 and the superstrate ($n_1 < n_2 > n_3$);
(ii) minimize optical loss by control of the guide thickness to monomolecular layer dimensions; and
(iii) render patterned guided wave structures of micron dimension by standard optical, x-ray, and electron beam lithographic procedures.

In one approach, considerable progress is being made utilizing deposition of disubstituted diacetylene monomolecular films onto solid substrates by the Langmuir-Blodgett technique(18,19). In this technique, a monomolecular layer of amphiphilic diacetylene monomer molecules is spread onto a air-water interface and compressed. As shown in Figure 15, typical pressure-area isotherms exhibit a characteristic rapid pressure rise as the spread diacetylene molecules are oriented upright to the air-water interface into a condensed phase(19). The condensed monolayer is transferred onto a vertically dipped solid substrate and multilayer films are systematically built up by sequential monolayer deposition thereby controlling film thickness at the monomolecular level.

Studies of the solid state photopolymerization of deposited multilayer films reveal that uv induced photopolymerization of disubstituted diacetylenes

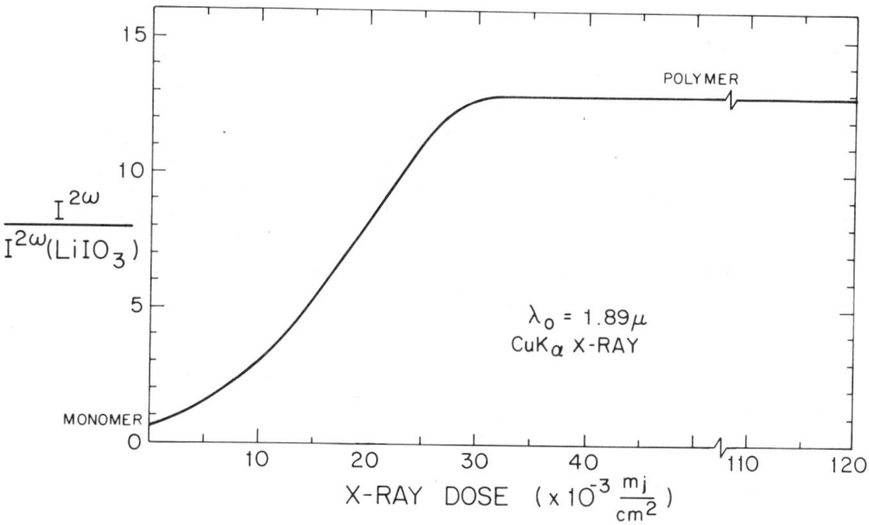

Figure 14. The second harmonic $I^{2\omega}$ *of NTDA microcrystals relative to the second harmonic intensity of lithum iodate ($LiIO_3$) powder* $I^{2\omega}$ *with increasing x-ray-induced polymerization. (Reproduced with permission from Ref. 7. Copyright 1980,* J. Opt. Soc.*)*

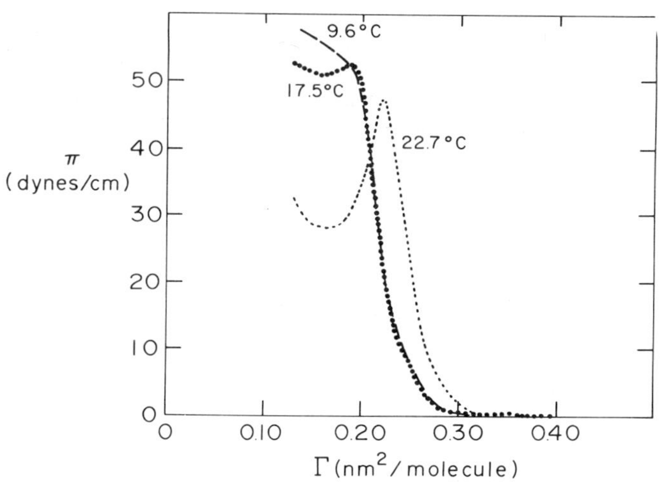

Figure 15. Pressure-area isotherms for spread diacetylene monolayer of cadmium pentacosa-10,12-diynide.

proceeds with quantum efficiencies in the range 10-100 depending on the particular monomer molecular structure (18). Separate studies of x-ray induced polymerization of film forming disubstituted diacetylenes have demonstrated quantum efficiencies as high as 10^{12}.(19) In contrast to uv photons, absorbed x-ray photons provide a natural gain mechanism by creating large numbers of secondary electrons and other photons that actually cause polymerization. An example of pentacosa-11,12-diynoic acid is shown in Figure 16.

High resolution patterns can be rendered in disubstituted diacetylene films by selective polymerization using standard uv, x-ray, and e-beam lithographic methods(19). An example of a 1 micron size test pattern of a developed diacetylene polymer film deposited on an oxidized silicon wafer by Langmuir-Blodget methods is shown in Figure 17. Such patterns are developed, and if desired, transferred to the underlying substrate by liquid reagents. However, it was recently found that such patterns can also be processed by dry plasma etching, which is important to avoiding polymer swelling, peeling and other undesirable pattern irregularities often introduced by using liquid reagents.

In addition to $\chi^{(2)}$ and $\chi^{(3)}$ processes in these diacetylene polymer guided wave structures, current studies are focused on recently developed film forming disubstituted diacetylenes exhibiting phase matched second harmonic generation such as DCNQDA (Figure 18)(20). In the quinoid ring, the π-electron system is substituted with the cyano (CN) electron acceptor and amine (-NH) donor groups. Development of these new structures is based on earlier theoretical calculations for the parent molecular unit DCNQDA(20). An exceptionally large value of -170 x 10^{-30} cm^5/esu at 1.06μ had been predicted resulting from charge correlations in the π-electron excited states and subsequently confirmed by DCSHG experiments on a related derivative(20).

Conclusion

In summary, we have briefly reviewed current research highlights from studies of second order nonlinear optical responses in organic and polymeric media. We have stressed how fundamental studies have led to microscopic understanding of important electronic states that comprise the origin of the large second order nonlinear responses in these

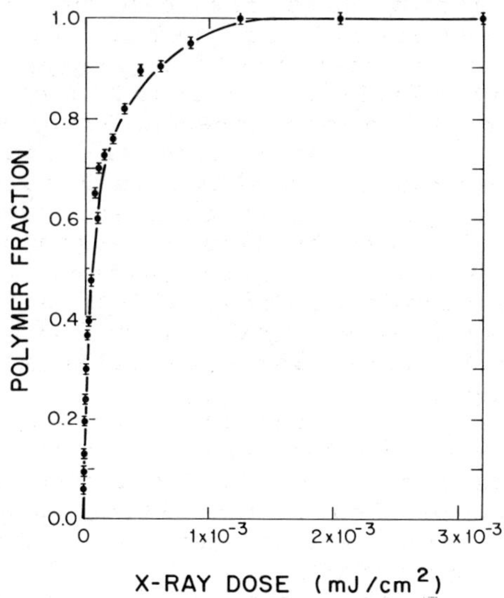

Figure 16. x-Ray induced polymerization of pentacosa-10,12-diynoic acid with quantum efficiency of 10^{-12} (CuK_α).

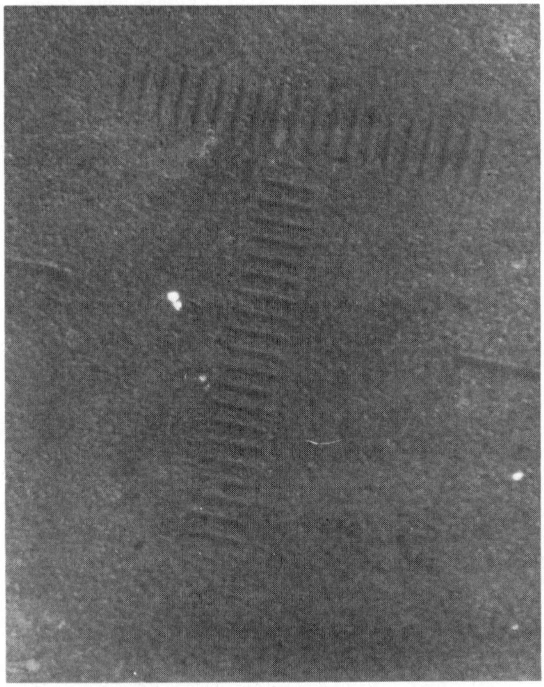

Figure 17. Photomicrograph of a UV lithographic T feature showing 1 μm-wide spaces and 2 μm-wide bars in the center of a 300 μm-wide square.

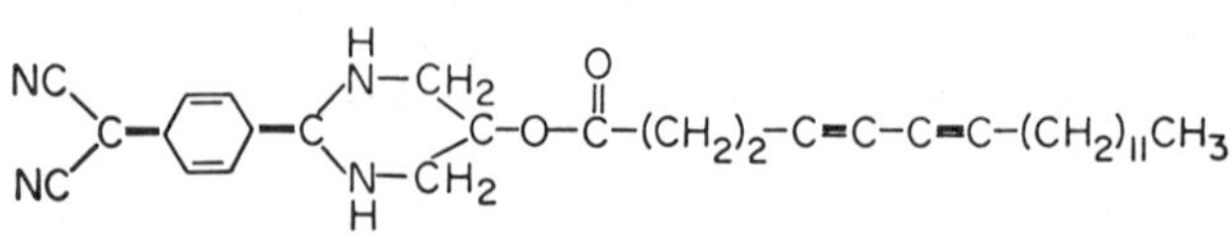

Figure 18. Quinoid structure of DCNDA. (20).

systems. Moreover, the feasibility of phase matched second harmonic generation in single crystal polymers and high resolution lithography for guided wave structures has been demonstrated with disubstituted diacetylene polymers. This multidisciplinary field has witnessed important advances both in theory and experiment. However, this promising field will remain in its initial phases without progress on the major problems of crystal growth and film fabrication of these materials. Finally this short review has centered on second order processes, but we anticipate similar progress in the understanding of important third order processes.

Acknowledgments

This work was supported by the Defence Advanced Research Projects Agency under No. DAAK-70-77-C-0045 (5-26502) and the National Science Foundation MRL program under grant No. DMR-7923647.

Literature Cited

1. B.F. Levine, C.G. Bethea, C.D. Thurmond, R.T. Lynch and J.L. Bernstein, J. Appl. Phys. 50, 2523 (1979).
2. G.F. Lipscomb, A.F. Garito and R. Narang, J. Chem. Phys. 75, 1509 (1981).
3. A.Carenco, J. Jerphagnon and A. Perigaud, J. Chem. Phys. 66, 3806 (1977).
4. J.L. Oudar and R. Hierle, J. Appl. Phys. 48, 2699 (1977).
5. J. Zyss, D.S. Chemla and J.F. Nicoud, J. Chem. Phys. 74, 4800 (1981).
6. J.A. Morrell, A.C. Albrecht, K.H. Levin and C.L. Tang, J. Chem. Phys. 71, 5063 (1979).
7. A.F. Garito, K.D. Singer, K. Hayes, G.F. Lipscomb, S.J. Lalama and K.N. Desai, J. Opt. Soc. 70, 1399 (1980).
8. A.F. Garito and K.D. Singer, Laser Focus 80, 59 (1982) and refrences therein.
9. K.D. Singer and A.F. Garito, J. Chem. Phys. 75, 3572 (1981).
10. B.F. Levine and C.G. Bethea, J. Chem. Phys. 63, 2666 (1975).
11. J.L. Oudar and H. LePerson, Opt. Comm. 15, 258 (1975); 18, 410 (1976).
12. C.C. Teng and A.F. Garito, Phys. Rev. Lett. 50, 350 (1983) and to be published.

13. S.J. Lalama and A.F. Garito, Phys. Rev. A20, 1179 (1979).
14. G. Wegner in Molecular Metals, W. Halfielded, Plenum (1979) and references therein.
15. A.R. McGhie, G.F. Lipscomb, A.F. Garito, K.N. Desai and P.S. Kalyanaraman, Makromol. Chem. 182, 965 (1981).
16. G.F. Lipscomb, A.F. Garito and T.S. Wei, Ferroelectric 23, 161 (1980).
17. C. Sauteret, J.-P. Hermann, R. Frey, F. Pradere, J. Ducuing, R.H. Baughman and R.R. Chance, Phys. Rev. Lett. 36, 956 (1976).
18. B. Tieke and G. Wegner, Makromol. Chem. C179, 1639 (1978).
19. J.E. Sohn, A.F. Garito, K.N. Desai, R.S. Narang, and M. Kuzyk, Makromol. Chem. 180, 2975 (1979); and also S.E. Rickert, A.F. Garito, K. Hayes, K.N. Desai and M.E. Filipkowski, (to be published).
20. S.J. Lalama, K.D. Singer, A.F. Garito and K.N. Desai, Appl. Phys. Lett. 39, 940 (1981) and to be published.

RECEIVED August 5, 1983

2

Design and Characterization of Molecular and Polymeric Nonlinear Optical Materials: Successes and Pitfalls

GERALD R. MEREDITH

Xerox Corporation, Webster Research Center, Webster, NY 14580

Over the past decade it has been learned that organic materials containing appropriately constituted or substituted conjugation systems may exhibit highly enhanced electronic nonlinear optical polarization responses (1-4). Since the microscopic second-order nonlinear hyperpolarizability tensor, β, is orientationally averaged to zero in random media (except the nonuseful pseudoscalar component (5)), the realization that elements of β could be so significantly enhanced developed in large part as result of studies of electric field induced second harmonic (EFISH) generation in molecular liquids (1-2). In the EFISH technique characteristics of β are sampled as a weak molecular alignment is induced by a static electric field and the subsequent frequency doubling of laser light is observed. Correlation of molecular charge transfer characteristics with β enhancement was shown and several highly nonlinear crystals were discovered based on this knowledge (6-8). However, a major limitation to achieving the enormous potential of molecular materials (e.g. crystals with $\chi^{(2)} \sim 10^{-5}$ esu (3) to be compared to $\chi^{(2)} \sim 10^{-9}$–10^{-7} esu of common nonlinear optical media) has been the requirement to achieve appropriate alignment of microscopic units. This requirement is to minimize orientational cancellation of the third rank polar β tensors whose contributions determine the macroscopic second-order electric dipole susceptibility, $\chi^{(2)}$. In nearly all cases, such alignments are achieved through packing forces in crystals with structures belonging to one of the noncentrosymmetric crystal classes. Unfortunately, details of these forces are currently too poorly understood to enable a molecular design approach for crystal structure and molecular alignment control.

In the first part of this paper several novel approaches to the alignment problem of second-order nonlinearity will be presented. The molecular information which guided and motivated these efforts, and others described in this symposium, derives from third-order nonlinear optical experimentation (EFISH and others). In the second

0097–6156/83/0233–0027$08.50/0

half of this paper inaccuracies, subtleties and pitfalls of these and other third-order experiments will be discussed. Besides β, third-order electronic nonlinear hyperpolarizabilities, γ^e, are also known to be enhanced in certain organics. Due to properties of polar fourth rank cartesian tensors, in contrast to the third rank polar $\chi^{(2)}$ tensors, alignment is a less significant factor in obtaining media displaying highly enhanced $\chi^{(3)}$ from molecules possessing large γ^e (4). This, coupled with predictions of recent molecular orbital calculations of γ^e (9), have led us to develop accurate techniques to reliably study these quantities. These techniques and results will also be discussed.

Methods and Evaluation of Second-Order Nonlinearity Preserving Alignments

Crystals. A necessary condition for $\chi^{(2)}$ to be nonvanishing is that the medium have a noncentrosymmetric microscopic organization. The most common such structures are crystals with space groups from the noncentrosymmetric classes. For molecular crystals to exhibit nonnegligible $\chi^{(2)}$ both the molecular structure and the unit cell structure must meet this condition. The former condition is met by the utilization of compounds satisfying conditions which enhance β. Recognizing the difficulty of predicting the nature of crystal unit cell structures, if one is interested simply in identifying potential new nonlinear media, as a lowest level strategy one might survey large numbers of crystals of these compounds with the hope of finding a few favorably aligned. A method which may be employed for this evaluation, as an alternative to time consuming single crystal characterization, is observation of second harmonic generation (SHG) from crystalline powders (10-11). We have screened several hundred compounds by SHG powder methods, as have several other research groups. These methods represent a very complex optical situation in which the observation of one intensity or a series of intensities as function of powder size or doubling wavelength is performed and as such are not quantitative, are subject to artifacts and must be approached with a specific set of expectations for their interpretation.

We recognize that there are applications in two- and three-dimensional waveguides (12,13) which do not have the same criteria of phase-matching as in simple crystals or that one may just as well be interested in screening these materials for the related electro-optic performance by the simple SHG powder method. (It has been shown for several organic materials that although the electro-optic and SHG $\chi^{(2)}$ tensors are in principle unequal, due to dispersion and due to the possible contribution of atomic and molecular distortions in the former, the predominance of the electronic contributions to $\chi^{(2)}$ in these materials causes these tensors to be nearly equal (14-16).) It is, therefore, worthwhile to know that limitation of powder particle sizes to less than a typical SHG coherence length allows more elements of the $\chi^{(2)}$ tensor to contribute significantly, and thus be sampled, in the SHG test than when larger particle sizes are used. (The coherence length is the distance over which the nonlinear polarization wave, $\underline{P}^{2\omega}=\chi^{(2)}\cdot\cdot\underline{E}^{\omega}\underline{E}^{\omega}$, responsible for SHG acquires a phase mismatch in temporal oscillation of magnitude π relative to that of the normally propagating light wave at frequency 2ω because of dispersion of refractive index and birefringence; effective interaction lengths caused by beam walk-off in birefringent crystals obviously are not appropriate in powders (17,18).) Larger particle size serves as a discriminator for single crystal phase-matching properties and the components of $\chi^{(2)}$ so associated (10,11). Further, to allow the screening of a wider range of materials, such as dye compounds, we have undertaken to monitor SHG appearing at 954 nm in response to the intense 1.91 μm wavelength light produced by stimulated Raman conversion of the 1.064 μm output of a Q-switched Nd^{3+}/YAG laser. Such low frequency light gives a test less dependent on the influences of electronic resonance on refractive index and nonlinearity dispersion than the more typical 1064→532 nm SHG tests. In keeping with these comments on generality and particle size, in our tests, particles have been typically in the approximate range 20 to 40 μm. This relatively large size still assures nondiscrimination by phase matching characteristics due to smaller index dispersion and longer wavelength in 1.91→0.954 μm SHG.

Strategies for discovering new organic nonlinear crystalline materials have been varied. As mentioned above, the most straightforward is a "shotgun" approach wherein many available materials are tested. A first level strategy which assures at least that the compounds will exhibit noncentrosymmetric structures is to test crystals of either enantiomorph of chiral compounds (7,19). It is our belief that this is a mathematically correct concept, but the physics of this situation does not contain specific intermolecular forces which assure favorable alignment of the portion of the molecules which possess the enhanced nonlinear polarizability. In a second strategy, employing some physical intuition into packing forces, it is hypothesized that a correlation between the dipolar and conjugation characteristics of nonlinearly enhanced molecules and the common occurrence of antiparallel alignments exists (8). We have attempted to overcome the structure determining influences of these factors using the general requirement of dense molecular packing in crystals (20) but employing less geometrically simple molecules created by asymmetrical addition of bulky substituents. While we have identified several materials displaying powder SHG comparable to the best reported organic compounds (6-8), space limitations dictate that we detail results of a third strategy. In this strategy some second component which is intended to modify or affect the packing of the nonlinear moiety occurs in the crystal.

Under this last strategy we have studied molecular salts chosen such that the cation carries β enhancing attributes. An experimental advantage of this approach is that by metathesis one can straighforwardly obtain new compounds and crystals with varied anions, thus eliminating the need for new synthesis with each compound modification one hopes will affect crystal structure and improve alignment. In molecular salts we expect that monopolar interactions will override the deleterious dipolar effects in determining the alignments of β enhanced species. Examination of structures of related cationic dye salts supports the validity of this idea. In the iodide salt of the trans-4′-dimethylamino-N-methyl-4-stilbazolium ion the cations stack in layers (as in smectic liquid crystal) in which they are tilted relative to the plane normal but stand parallel in a polar sense (21). This structure can be rationalized by the need first to maximize the Madelung energy through the arrangement that the pyridinium rings and iodide ions be in close proximity and second to maximize dispersion energy through parallel side-by-side stilbazolium packing. Unfortunately, further increase of Madelung energy results from arrangement of an inverted layer of parallel cations accompanied by counterions adjacent to the first, doubling the size of the ionic region but causing a centrosymmetric structure. The principles of maximizing coulombic and dispersion stabilization energies can also be used to partially explain the structure of the (+)-camphor-10-sulfonate salt of the trans-4′-hydroxy-N-methyl-4-stilbazolium ion (19), which is a protonated analogue of a simple merocyanine dye class exhibiting the largest reported values of β (4,22). In this salt layers of distinct type (aliphatic-ionic-aromatic-ionic-aliphatic-etc.) exist as can be seen in Figure 1. The unfortunate nonlinearity cancelling pseudo-antiparallel orientation of the cations (as pseudo-inversion dimers) can be partially attributed in this case to the large size of the anion (a volume packing argument) and to the stabilizing influence of hydrogen bonding between the hydroxyl and sulfonate groups.

To avoid the apparently adverse effects of hydrogen bonding with the last cation mentioned we have concentrated on salts of the dimethylamine cation. Also in consideration of the importance of an anion bulk packing factor we have tried several size and shape anions, finding that members of a class of crystals involving medium small, largely pseudo-tetrahedral anions typically exhibit large powder SHG - up to an order of magnitude greater than the best

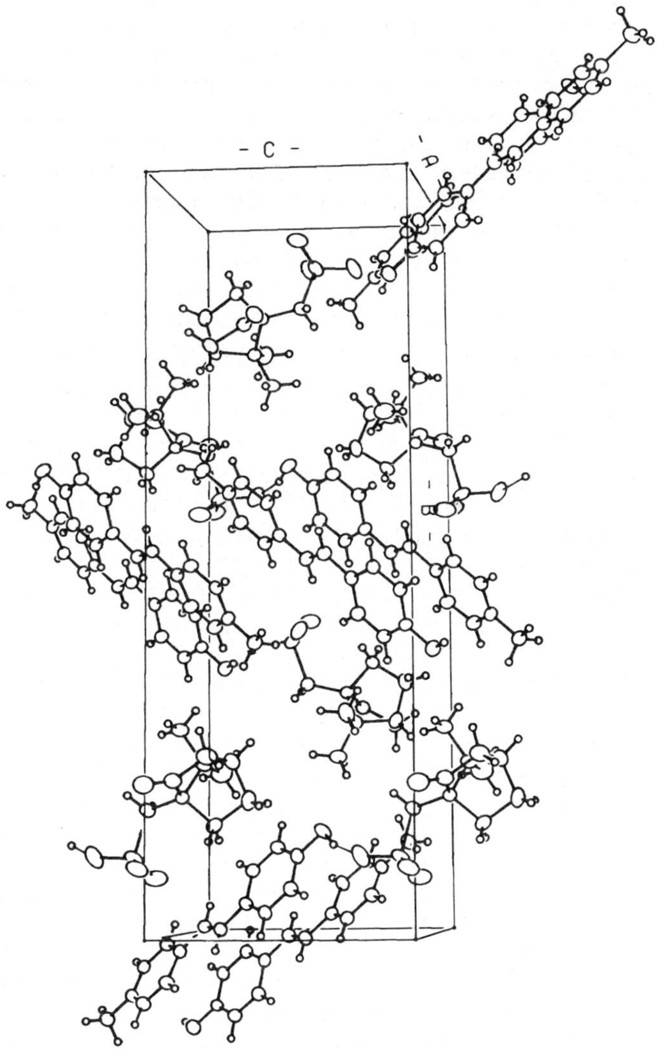

Figure 1. Crystal structure of trans-*4′-hydroxy-*N-*methyl-4-stilbazolium (+)-camphor-10-sulfonate.*

organics in the literature. Table I contains data for some salts to be compared to three known materials, MNA having the largest reported $\chi^{(2)}$ among organic crystals (6). Results for the two salts discussed in the last paragraph are consistent with their structures. The sensitivity of the structure of the trans-4′-dimethylamino-N-methyl-4-stilbazolium iodide salt to the monopolar interactions is evidenced by the substantial SHG value observed in the trans-4′-dimethylamino-N-ethyl-4-stilbazolium iodide salt; the replacement of a methyl group on the charged pyridinium ring by an ethyl group apparently reduces Madelung energy and destabilizes the centrosymmetric structure relative to some other which is noncentrosymmetric.

Table I. Relative SHG Powder Intensities.

Crystal		Harmonic Intensity
α-quartz		0.008
metanitroaniline (mNA)		1
2-methyl-4-nitroaniline (MNA)		3
trans-4′-hydroxy-N-methyl-4-stilbazolium (+)-camphor-10-sufonate		0.007
trans-4′-dimethylamino-N-ethyl-4-stilbazolium iodide		0.7
trans-4′-dimethylamino-N-methyl-4-stilbazolium X		
X^- =	I^-	0
	IO_3^-	0.01
	NO_3^-	0.5
	cinnamate	1.5
	ClO_4^-	5
	BF_4^-	10
	ReO_4^-	18
	$CH_3SO_4^-$	30

X-ray structural analysis of the methylsulfate compound indicates the orthorhombic crystal unit cell contains two translationally inequivalent cations positioned on mirror planes and tilted at 34° relative to the two-fold screw (c) axis (23). This is a compromise orientation for simultaneously, rather than individually, maximizing $\chi^{(2)}_{ccc}$ and $\chi^{(2)}_{P\{bbc\}}$ in this polar structure. This structure is therefore consistent with the extremely large SHG intensity reported in Table 1 while, also consistently, preliminary x-ray data show the perrhenate and tetrafluoroborate salts to be isostructural (23). Details of the packing considerations in this class have not yet been analyzed. Attempts to grow large single crystals to be used for $\chi^{(2)}$ tensor characterization are in progress. While an oriented molecule analysis is not completely rigorous due to complexities of the local field effects, one would expect to be approximately correct in predicting the properties of $\chi^{(2)}$ through such an exercise. Unfortunately the characteristics of β in the trans-4'-dimethylamino-N-methyl-4-stilbazolium ion cannot be determined by the EFISH technique due to the inability of ionic solutions to support easily characterized static electric fields. However, by a method of analysis of optical third harmonic generation (THG) in solutions utilizing cascading of molecular second-order nonlinear responses, we have recently inferred that the second-order nonlinearity of this cation is nearly equal to that of 4-dimethylamino-4'-nitrostilbene (DANS). DANS has been characterized by EFISH to be an order of magnitude more nonlinear than the best single ring species such as p-nitroaniline (pNA) and MNA (2,6). In the orientation of the methylsulfate salt, if one reasonably assumes the cation's second-order response to occur predominantly in the β_{zzz} tensor element (z being the cation long axis), trigonometry predicts that ~55% and ~70% of the maximum attainable $\chi^{(2)}{}_{iii}$ or $\chi^{(2)}{}_{P\{ijj\}}$ in this or any other structure occur. It is therefore expected that the larger elements of the $\chi^{(2)}$ tensor of these salts will prove to be substantially greater than those achieved in crystals of pNA and MNA related molecules except in particular resonance enhancement frequency regimes.

Quasicrystals. As described in an earlier paper of this symposium there is an interesting photochemically initiated noncentrosymmetric ordering of dipolar species derived from solutions of certain spiropyran compounds dissolved in aliphatic solvents (24). This ordering is driven by sudden, drastic change of solute character. Since the photochemically modified spiropyran has a merocyanine form which is expected to display second-order nonlinearity, the net effect of the spontaneous ordering is also expected to create a nonvanishing macroscopic second-order nonlinearity. The process of alignment is understood to proceed through the following sequence: the initial photoprocess creates at least two species containing differing numbers of merocyanine and spiropyran molecules. Stacked aggregates of one entity form and coalesce into microcrystallites with diameters approximately 40 nm. These are covered with amorphous material deriving from another initial species, forming globules which appear in discrete sizes indexed by generation number (there being a time sequence for their creation governed by the changing ratio of concentrations of the globule forming species remaining in solution). If formed in an electric field, globules fuse in a string-of-beads morphology. This process is attributed to energetic limitations on globule contact approach paths for highly dipolar particles aligned by the field.

We have used SHG to confirm and probe these structures. Quantitative characterization is complicated since the globules of a sample have correlated but not perfect positioning and they are smaller but not substantially smaller than the wavelengths of light employed. Thus determination of far-field harmonic radiation patterns based on SHG response to a laser beam is not straightforward. We have found though that various schemes of index matching and EFISH type experiments allow us to observe SHG, comment on its origin, to observe modifications of structure in a static field and to follow dynamics of globule modification (25,26). The reader is referred to the accompanying paper on this topic.

Molecularly Doped Liquid Crystalline Polymer. Regarding the infrequent occurrence of crystals of highly nonlinear molecules showing substantial alignment, as monitored by SHG, one might attribute this to antiparallel, dimer-like alignment favored by more symmetric or simple molecular shapes, enhancement of dispersion forces and minimization of dipole-dipole interaction energies (8). While large permanent dipole moments appear to be a hinderance to appropriate alignment in crystals, they allow that substantial macroscopic nonlinearity may be achieved when the more highly nonlinear species are aligned in fluid media by a large static electric field. If a liquid crystal solvent in which the nonlinear molecule exhibits a large alignment correlation is employed, the alignment portion of the resultant susceptibility may be enhanced by up to a factor of five over that in isotropic solvents. Alternatively, by an appropriate sequence of temperature and field treatment, the alignment may be induced in a low viscosity state and become permanent in a higher viscosity state of a polymer (27), thus producing a material with nonvanishing $\chi^{(2)}$. Such an alignment is, of course, unstable and would gradually disappear if not for the possibility of strain and impurity charge compensation of the polarization density field. By combining the liquid crystalline and polymeric attributes both gains are achieved, but an additional advantage is obtained. The high cooperative light scattering and attenuation coefficient which is a tremendous disadvantage for the application of typical liquid crystals in nonlinear optical devices is reduced by the large increase in internal viscosity. A suitable material has been described in an earlier paper of this symposium. DANS (4-dimethylamino-4′-nitrostilbene), which has one of the largest β's known and which displays a large $S^{(2)}$ order parameter in that material, was used at 2% dopant concentration. SHG was used to characterize this material (28) and the reader is referred to the paper in this volume describing the findings.

Characterizations by Third-Order Nonlinear Optical Experimentation

Except through the study of linear and nonlinear optical properties of molecular crystals, methods to determine the nature of β require evaluation of appropriate $\chi^{(3)}$ characteristics of molecular ensembles in conjunction with some type of modeling of the manner in which molecular nonlinearity contributes to the bulk susceptibility. We have considered the two aspects of this task and have found that due to the complexity of the considerations there are many pitfalls. In this second part of the paper discussions are presented which deal with light propagation effects affecting $\chi^{(3)}$ determinations, which will show, among other things, the necessity for newly developed techniques to measure $\chi^{(3)}$ by optical third harmonic generation (THG). Subsequently comments about molecular interpretations of $\chi^{(3)}$ are made.

Nonintuitive Light Propagation Effects In Third-Order Experiments. One of the first tasks for a chemist desiring to quantify second- and third-order optical nonlinear polarizability is to gain an appreciation of the quantitative manifestations of macroscopic optical nonlinearity. As will be shown this has been a problem as well for established workers in the field. We will present pictures which hopefully will make these situations more physically obvious.

Two straightforward third-order nonlinear optical phenomena which are used to characterize β and γ are EFISH (1,2,4) and THG (3) (or their multicolor analogues (29)). Using classical physics descriptions, as source of these phenomena there exist nonlinear polarization waves oscillating at frequencies 2ω or 3ω (30),

$$\underline{P}^{2\omega}(\underline{r},t) = 3\chi^{(3)}(\underline{r})\cdots\underline{E}^{\omega}(\underline{r},t)\underline{E}^{\omega}(\underline{r},t)\underline{E}^{0}(\underline{r},t)$$

$$\underline{P}^{3\omega}(\underline{r},t) = \chi^{(3)}(\underline{r})\cdots\underline{E}^{\omega}(\underline{r},t)\underline{E}^{\omega}(\underline{r},t)\underline{E}^{\omega}(\underline{r},t)$$

induced by components of the total electric field which are usually nearly monochromatic, linearly polarized travelling waves having some slowly time varying amplitudes (the sources being pulsed high voltage across electrodes and/or high peak power pulsed lasers). Spatial variations of $\underline{E}$ and $\underline{P}$ will exist between the different media and across the profile of laser beams used in real experiments. It is not easy in general to predict the form, magnitude or dependencies of the total harmonic waves "broadcast" from the oscillating polarization densities resulting from focused laser beams propagating through different materials and possibly through the fringing fields of the electrodes used to establish the zero frequency electric field for EFISH. Some surprising propagation phenomena are: 1) EFISH is observed from gases placed between charged capacitor plates yet none is seen from liquids unless two dissimilar materials (e.g. liquid and a glass plate) form an optical interface, and the resulting signal depends on the differences in response of the two materials at the interface; 2) true THG is not observed from normally dispersive material when a gaussian laser beam is focused through it, but here also THG is observable if an interface is created, the magnitude again depending on the difference of nonlinear responses of the two materials at that interface; and 3) SHG by a focused gaussian beam in a finite noncentrosymmetric crystal is sometimes maximum when the minimum beam waist falls at a crystal face. Two pictorial models are useful in understanding or describing these and other related phenomena: considerations of the coherent field from amplitude and phase modulated dipole arrays, and vibration diagrams.

These problems of macroscopic manifestation of the nonlinearity hinge on aspects of energy transfer through the nonlinear polarization waves among the normal (linear) electromagnetic waves of the media. Usually in characterization experiments conversions are negligibly small so that nonlinear modification of the laser fields may be neglected and one may concentrate on variation of the harmonic field. The local relationship between the oscillating nonlinear polarization density and the linear oscillating electric field determine the result. As in linear optics (31) where the in-phase component of polarization density is associated with retardation (refractive index) and the out-of-phase component with attenuation (absorption coefficient) or gain (associated with a π phase alteration of $\underline{P}$ relative to the loss case, as in lasers or optical amplifiers), the variation of phase of $\underline{P}^{n\omega}$ relative to that of $\underline{E}^{n\omega}$ governs whether the harmonic wave is amplified or diminished in amplitude (and whether a change of phase occurs). It is largely the effects of variation of the phase-mismatch between these waves, integrated along the optical path, which is important in determining the magnitude of detected harmonic light. As will be done below, it is preferable in attempting to describe the integrated effect to factor out the rapid spatial variation of the phase of the linear electromagnetic wave and picture only this relative phase variation along the light rays. While wavelengths of light are on the scale of microns, distances along which these variations occur vary from a few microns to meters.

Dipole Array Model. In Figure 2 several illustrative cases are depicted. The magnitude of the component of discrete dipoles ($\underline{P}^{n\omega}$ integrated over some small volume) oscillating out-of-phase with $\underline{E}^{n\omega}$ along one ray is depicted. (Details of transverse structure are ignored, it being assumed that sufficient breadth exist; that diffraction of the resultant electric wave is negligible.) The question of resultant energy transfer is the same at optical or radio frequencies, the latter being more intuitive. In Figure 2a r oscillating dipoles (antennae) broadcast "in-phase with one another". The downstream field is $r(\underline{e}^{n\omega}p)$ where $\underline{e}^{n\omega}$ is the field of one dipole element (really one transverse plane of dipoles) divided by its amplitude p. Since intensity is proportional to the square of electric field, $I^{n\omega} \propto (rp)^2$. This illustrates that in phase-matched interactions intensity varies as the interaction length times susceptibility times $(I^{\omega})^{n/2}$ total quantity squared. Relating the discrete model to continuous media this is $I^{n\omega} \propto (rp)^2 \propto (l\chi(E^{\omega})^n)^2 \propto (l\chi)^2(I^{\omega})^n$. In the region of the dipoles variation of the harmonic field is simple. Since the phase is depicted relative to a forward travelling wave, the electric field variation in that region depends on details of a backward travelling wave. The corresponding picture of out-of-phase dipoles for the backward wave would vary sinusoidally, but with wavelength half that of the linear electromagnetic wave - a situation incapable of broadcasting more than a negligible field at optical frequencies where the wavelength is very short compared to array (sample) size. Therefore $\underline{E}^{n\omega}$ grows linearly as the dipole array is traversed in the forward direction. The latter conclusion is not necessarily so at radio wavelengths (or for the case of a few stacks of quasicrystals at optical frequencies).

Figure 2b depicts a case where $\underline{P}^{n\omega}$ possesses a linearly increasing phase deviation relative to the linear forward travelling $\underline{E}^{n\omega}$ wave, as occurs for harmonic generation when refractive indices of fundamental and harmonic waves are not equal. Downward pointing $\underline{P}^{n\omega}$ indicate phase relationships which extract energy from the $\underline{E}^{n\omega}$ wave. Consideration of "broadcasting" in this situation shows that $\underline{E}^{n\omega}$ grows and shrinks periodically with the period of the phase-mismatch as function of z in the region of the dipoles. This approximately depicts the behavior of non-phase-matched harmonic generation as function of position, z (or crystal thickness). Also an important point which can be verified with a little thought is that the amplitude of the sinusoidal spatial variation of $\underline{E}^{n\omega}$ is proportional to pl_c, the magnitude of the dipoles times the effective broadcasting distance before phase reversal. One can readily see through these pictures the origin of the standard relation $I^{n\omega} \propto (\chi l_c)^2(I^{\omega})^n \sin^2(\pi z/2l_c)$, where l_c is the coherence length (distance for accrual of a π phase mismatch).

The following cases are somewhat more subtle. Figure 2c depicts a slowly varying amplitude $\underline{P}^{n\omega}$ with a very slowly varying phase-

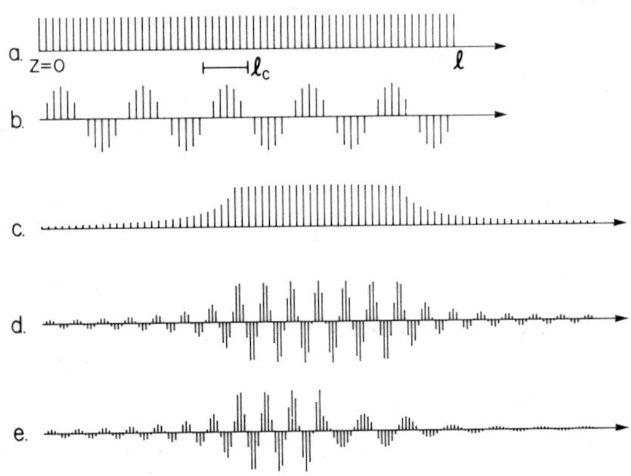

Figure 2. Dipole array pictures of some situations in nonlinear optical harmonic generation. (See text for further discussion.)

mismatch. This corresponds to EFISH in gases where the spatial extent of the typical electrode field is smaller than the coherence length for SHG (consider $l_c=\pi c/2\omega\{n_\omega-n_{2\omega}\}$). A substantial EFISH signal may result since little phase deviation exists. In contrast Figure 2d depicts a similar situation, but in which a much more rapidly varying phase-mismatch appears. This might correspond to EFISH in condensed phase if hundreds rather than ~10 cycles of phase-mismatch occured in the electrode region. Absolute dispersion of refractive indices being substantially greater in condensed phases than in gases causes correspondingly shorter l_c. A totally different picture results: an insignificant harmonic field will be "broadcast" to the large z region, but nonnegligible $\underline{E}^{2\omega}$ will appear in the region of large E^0 between the electrodes. The amplitude there is proportional to $l_c p \propto l_c \chi (E^\omega)^2 E^0(z) \propto l_c \chi I^\omega E^0(z)/n_\omega$ where $E^0(z)$ is the amplitude of the static field at z and where the relation $I^\omega \propto n_\omega |\underline{E}^\omega|^2$ has been used. For this reason, to observe EFISH in liquids a sharp boundary is usually created in the region of constant static field by insertion of at least one glass slab. This situation is depicted by Figure 2e. The resultant amplitude of $\underline{E}^{2\omega}$ observed at large z would then be approximately proportional to $\{(\chi l_c/n_\omega)_1-(\chi l_c/n_\omega)_2\} I^\omega E^0(z')$ with z′ being the location of the interface. (Obviously if zero magnitude $\underline{E}^{2\omega}$ occurs at large z after passage through the electrodes with material 2 totally filling the space as in Figure 2d and if the field at point z′ in that case is proportional to $(\chi l_c/n_\omega)_2 I^\omega E^0(z')$, then the field generated by the dipole array from point z′ to large z must be similarly proportional to $-(\chi l_c/n_\omega)_2 I^\omega E^0(z')$. Therefore, $\underline{E}^{2\omega}$ at large z in the two media case is proportional to the sum of fields generated from each medium, which is proportional to the difference quoted, if small corrections for interface physics are ignored.)

<u>Vibration Diagram Method</u>. In actuality the last cases above are not described accurately by this dipole array model because actual phases of the electric fields are significantly altered from those of linear waves. (A more realistic, but complex model is to consider amplitude and phase characteristics of the oscillating vertically polarized component of electric field resulting from rotation of a line of transverse dipoles of equal magnitude but rotated relative to each other along the line such that their vertical components at some reference time are depicted by Figure 2.) For this reason and to handle details of focused laser beams one must resort to a more mathematically based description. Fortunately, numerical

calculations describing propagation in several important nonlinear optical situations have been performed and summarized as vibration diagrams (32). Since EFISH and THG experiments have aspects which may be described by them, we've developed some interface nonlinear optical transmission algorithms to allow their use in our task of understanding pitfalls in materials characterization (33-35). Before discussing them we will review the vibration diagram method.

A vibration diagram is a graphical method which may be used to perform the integration of phase and amplitude effects of nonlinear polarization density along a ray, possibly as summary of an integration across the beam's transverse structure (32). Basically, at each point the nonlinear polarization density provides some differential broadcasting element with some amplitude function and phase relative to a hypothetical linear wave. A curve is then generated as the differential elements are integrated in the plane of the diagram. Tilt relative to preceding elements is the means of accumulation of phase-mismatch. The resultant $\underline{E}^{n\omega}$ at some point z along the optical path may be determined from the corresponding point on the resultant diagram: the distance of that diagram point from the origin point represents the amplitude of $\underline{E}^{n\omega}$ while the tilt of the connecting line segment depicts the phase relationship of $\underline{E}^{n\omega}$ relative to the reference linear wave.

Figure 3 contains the same examples as Figure 2. In these and the following diagrams z is a linear coordinate indicating distance along the optical path, ξ is a similar linear coordinate normalized for a gaussian (laser) beam such that $-1 \leq \xi \leq 1$ defines the near-field Rayleigh length, and ψ is an angular phase-mismatch parameter to be used where variation of phase-mismatch is very rapid over relatively small distances compared to the total optical situation, as when a small condensed phase portion is inserted into a gas phase diagram. At this point the reader is urged to interpret each of these diagrams with reference to the discussion above of Figure 2, remembering that the field magnitude at any point in the diagram (associated not necessarily through a linear relationship with the corresponding physical location in the real space of the experiment) is proportional to its distance from the z=0 point. One can see that indeed variation of phase is an important aspect to the resultant $\underline{E}^{n\omega}$ in cases b, d and e. Also one sees that the magnitude of fields generated depend on the size of the resulting diagrams. The latter obviously scale with the size of the nonlinear $\underline{P}^{n\omega}$ which, of course, scales directly with the nonlinear χ and the n/2 power of the laser power. Also the curved diagrams scale inversely with the rate at which the angular variation (phase-mismatch) accumulates, which is by definition inversely proportional to a coherence length.

Results of some of the numerical integrations of Ward and New (32) are also shown in Figure 4. Figure 4a depicts index-matched (that is, phase-matched for plane-waves) THG where a focused gaussian fundamental beam induces $\underline{P}^{3\omega}(\underline{r})$ which broadcasts into a corresponding confocal parameter harmonic gaussian mode. In this case, although nonzero $\underline{E}^{3\omega}$ occurs in the midregion, no resultant THG is created. The execution of a full 2π phase deviation can be understood from the fact that a gaussian wave accrues π less phase than a plane wave from the $-\infty$ to $+\infty$ limits. Then $\underline{P}^{3\omega}\propto(E^{\omega})^3$ eventually lags a plane wave by 3π while the 3ω reference gaussian lags by π, resulting in a relative 2π deviation in the diagram. In fact, for normally dispersive media ($n_{\omega}<n_{3\omega}$) no THG results, as Figure 4b depicts for the case $b\Delta k\simeq-0.3$ (b is the confocal parameter of the gaussian and Δk is the linear plane-wave, wave-vector mismatch, $\Delta k=3\omega(n_{\omega}-n_{3\omega})/c$). One can see that $b\Delta k>0$ results in nonvanishing net THG because only for $\Delta k=0$ does the exact 2π phase deviation relative to the reference 3ω gaussian required for the curve to close on itself without crossing accrue. Positive Δk opposes the loss of phase due to focused gaussian character, resulting in less curvature and open diagrams. Negative Δk increases curvature as in Figure 4b. Those diagrams are symmetric about a vertical line which is normal to the curves at $\xi=0$ and all close on themselves. Furthermore, Ward and New argued that since any focused laser beam can be decomposed into various gaussian modes and since their results hold for arbitrary gaussians, no THG should be observed from focused beams traversing normally dispersive media. As quoted above as second example of nonintuitive propagation effects, one could observe THG by creating an interface between two different media, as shown in Figure 4c, with the resultant THG then depending on the differences of nonlinear responses at the interface which in this case is positioned at the $\xi=0$ plane. A final example of the Ward and New calculations is the case of SHG with focused gaussian beams. Figure 4d shows an example. The diagram is broken because it is infinite. One can see, though, for finite size crystals, by taking the appropriate section of the diagram, as stated above as third nonituitive example, maximum SHG results if one positions the beam waist at one of the faces of the nonlinear crystal (provided the crystal size is larger than a few confocal parameters). Discussion of the $b\Delta k\neq0$ cases are similar to those of THG. This can pictorially explain both the best position of focus phenomenon and the fact that maximum conversion occurs for $b\Delta k>0$ (32,36).

To use these diagrams to describe real experiments where the beam passes through several media one must first consider the optical physics at interfaces between them. By simply requiring that

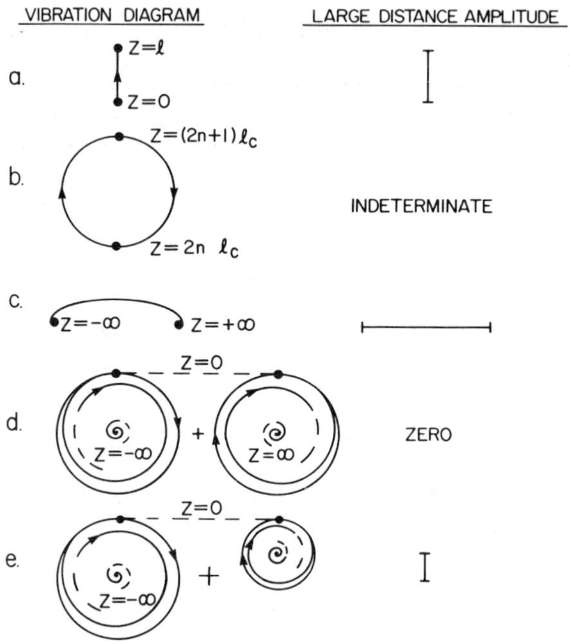

Figure 3. Vibration diagrams of some situations in nonlinear optical harmonic generation. (See text for further discussion.)

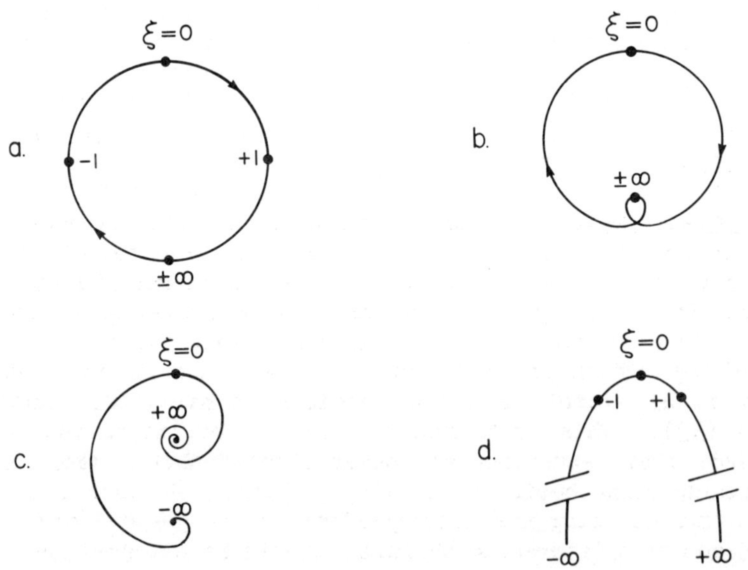

Figure 4. Vibration diagrams of some situations in nonlinear optical harmonic generation. (See text for further discussion.)

transverse electric and magnetic fields be continuous across an interface between media a and b and normal to the wave vector (assuming no reflected waves from other interfaces and only one nonlinear polarization wave), one can show (35) that the electric field oscillating at some chosen frequency at that interface is

$$\underline{E}(z_{ab}) = [2n_b/(n_a+n_b)]\,[\underline{E}^L{}_a + \underline{E}^{NL}{}_a] + [4\pi/(n_a+n_b)]\,[\{\underline{P}^{NL}/(n+n^{NL})\}_a - \{\underline{P}^{NL}/(n+n^{NL})\}_b]$$

where $\underline{E}^L$ is the component of $\underline{E}$ which propagates as the usual linear waves, $\underline{E}^{NL}$ is a component of $\underline{E}$ due to the presence of the nonlinear polarization wave $\underline{P}^{NL}$, and n^{NL} is a refractive index describing the propagating waves $\underline{E}^{NL}$ and $\underline{P}^{NL}$. $\underline{E}^L{}_a + \underline{E}^{NL}{}_a$ represents the resultant $\underline{E}_a$ due to the vibration diagram (and any injected linear wave) propagated up to the interface through medium a. The first term describes Fresnel transmission of light across the dielectric interface into medium b which is accompanied by the generation of some reflected waves in medium a. The second term is a component of electric field generated because of the discontinuity in $\underline{P}^{NL}$ between the media which is usually small enough to be neglected. Consequently, to splice together vibration diagrams $\underline{E}_a$ must be corrected by the transmission factor $2n_b/(n_a+n_b)$ and spliced to the diagram of medium b. Since this is a renormalization of distance from the beginning point of the vibration diagram, (a uniform expansion or contraction of the plane of the vibration diagram) it needn't be pictured but merely remembered when interpreting the resultant diagram. One must also remember that the linear waves which establish the magnitude of $\underline{P}^{NL}$ through $\chi^{(n)}$ must also be renormalized (vide infra).

Experimental Pitfalls. Several types of systematic inaccuracies in nonlinear optical susceptibility characterization techniques have appeared in the literature due to incomplete analysis of propagation effects. It is believed that use of the above models make them more obvious. Some examples are described in this section.

THG by condensed phase materials must be done in vacuum, otherwise, as example 2 above predicts, substantial inaccuracy results (33). This is a subtle effect. One might incorrectly concluded from experimental observations that there is no significant contribution from air. Figures 4a and b and the discussion concerning them above predict that no THG will be observed after focusing a laser beam through air. We have looked for such a signal under experimental conditions used for condensed phase THG studies and have indeed seen no signal when only air fills the optical path between focusing and collection lenses. However, the error results from the fact that, as Figures 4a and b predict, in the $\xi \simeq 0$ region there is nonvanishing $\underline{E}^{3\omega}$. The error is compounded since for parallel linear polarization the magnitude of $\underline{E}^{3\omega}$ due to the air in that region is proportional to $\chi^{(3)} l_c$ or to $\chi^{(3)}/[\chi^{(1)}_{3\omega} - \chi^{(1)}_{\omega}]$ which is a very weakly density dependent quantity. Therefore, since this is true of any material placed in the beam, gases establish THG fields comparable to many condensed phase materials. To see the effects of these fields on a typical experiment consider the effect of splicing air contributions onto the vibration diagram of some thin wedge of condensed phase material. In the absence of air Figure 5a would apply. The magnitude and phase of the resultant harmonic depends on the resultant phase-deviation upon passage through the material, that is, on the final value of ψ (which is one reason why a wedge is usually translated across the beam, allowing unambiguous observation and characterization of the entire nonlinear "Maker" fringe pattern and intensity). In Figure 5b, assuming the wedge thickness to be small relative to the confocal parameter of focusing, $-\infty < \xi < 0$ and $0 < \xi < \infty$ curves appropriate to air have been added to the wedge diagram. Remembering that the resultant $\underline{E}^{3\omega}$ is determined by the distance from the $\xi = -\infty$ to the $\xi = \infty$ point (for the particular value of ψ corresponding to the path through the wedge) the set of all possible $\xi = \infty$ points is displayed as the dashed curve. This latter curve is a circle, just as for the wedge alone, with a very small displacement from the $\xi = -\infty$ point due to interface transmission effects. The latter are difficult to observe and one might conclude from the apparently correct wedge fringing behavior and the observation of no interfering air THG upon removal of the wedge that all is correct. We've shown, however, that diagram 5b is indeed the case, since when the experiment is performed in an evacuable chamber, the resultant $I^{3\omega}$ (THG of a 1.91 μm wavelength fundamental) with glass or fused silica wedges may be increased by a factor of three to four on removing the air (33)! Comparisons of $\chi^{(3)}$ among materials via THG when this effect is ignored cannot, therefore, be correct.

Similarly one must be cautious in other third-order experiments. While EFISH of condensed phase materials looks like simple SHG if the sample is totally confined to the region of constant field between the electrodes, there is an important difference between this case and the more frequently performed and described study of frequency doubling crystals (37). In the latter case nonvanishing $\underline{P}^{2\omega}$ occurs only in the crystal where $\chi^{(2)} \neq 0$. However, in EFISH $\underline{P}^{2\omega}$ also occurs in air due to its nonvanishing $\chi^{(3)}$. Contributions to $\underline{E}^{2\omega}$ upstream and downstream from the sample must

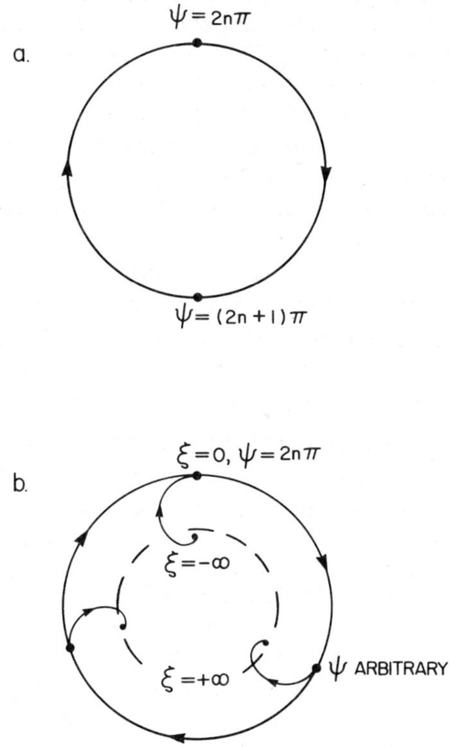

Figure 5. Vibration diagrams depicting the effect of air on optical third harmonic generation from a thin solid. (See *text for further discussion.)*

therefore be considered. Vibration diagrams similar to Figures 5a and b result but with the air contributions being more critically determined by details of focusing in relation to the fringing fields of the electrodes. This effect has not been previously appreciated for the determination of the EFISH nonlinearity of glass (2,38). Since that quantity is required to allow extraction of liquid $\chi^{(3)}$ in the usual glass-liquid-glass arrangement, one must be sceptical of claims of high accuracy when the effects of this systematic error are not included in the analysis. In our demonstration of this phenomenon (34) it was seen that a 7% error would have occurred for a λ=1.064 μm fundamental in the specific electrode and focusing geometry employed. In subsequent independent work comparable magnitude deviation was observed (39).

In an earlier section we noted that the field strength of the laser beam responsible for generating harmonics differs in various media. This relates to variable distribution of energy between electric and polarization density waves (31). Even if intensities were equal in two media, this and the effect of differing speeds of light can cause substantially different electric field strengths which is particularly important if they induce nonlinear polarization. An example of the subtlety of this factor is the case of EFISH in the common glass-liquid-glass arrangement. Although reflection losses between these media are typically less than 1%, electric fields may differ by several percent at comparable positions in the laser beam. Since the effect is amplified through the nonlinear response, inaccuracies greater than 10% in $\chi^{(3)}$ determinations may result from its neglect with normal organic liquids. Unfortunately this systematic error continues through the use of early EFISH formulas in which this factor has been overlooked (1,40). This has important implications in relation to studies of solutions of molecules in differing solvents.

There is another propagation related effect which has caused technique error in the past, but is not readily analyzed by the above models because it involves the nonlinear generation of several frequencies simultaneously. As example consider that in noncentrosymmetric solids it is usually the case that second-order allowed SHG occurs simultaneously with direct third-order THG (33,41). The additional allowed second-order frequency mixing, $2\omega+\omega\rightarrow3\omega$, called cascading (when occurring not due to the provision of both summing frequencies by the experimentalist), causes interferences in $\chi^{(3)}$ determinations. This effect, in addition to the neglect of atmospheric effects, invalidates the original α-quartz experiment (42) which was subsequently used for THG $\chi^{(3)}$ calibrations (4)! Recognition of this effect is particularly important since a portion of the cascading, when allowed by nonvanishing $\chi^{(2)}$ tensor elements, always generates a component of nonlinear polarization with identical frequency and wave vector as the direct third-order process under study. The magnitude of the cascaded contribution may be arbitrarily large, depending on details of $\chi^{(2)}$, $\chi^{(1)}$ and the experimental configuration. Because of the relative complexity of the formalism the reader is referred to other works for fuller exposition of these effects (33,41,43).

A significant problem in absolute susceptibility measurements is that of modal interferences when non-single-mode lasers are used to establish nonlinear polarization densities (37). Frequently the problens of multimodal lasers for susceptibility determinations are bypassed through the use of relative measurements for which identical behaviors occur. Alternatively, an internal microscopic interference is employed, such as the use of Raman interferences in $\chi^{(3)}$ characterization by CARS or RIKES, although those considerations are not necessarily simple or unambiguous (44-46). A case we have seen which is apparently significantly affected by multimodal sources is cascading in THG. Initially it was believed that the multimode associated ambiguity which arises when the final electric fields are added and squared to determine intensity would be unimportant. Here the "purely bound wave" cascaded contributions are created and summed in a similar manner to the direct THG response and the "intermediate free wave" contribution to $\underline{P}^{3\omega}$ is established over a distance short compared to the spectrally determined coherence length. However, comparison of nonlinearities of various materials determined relatively by totally third-order THG experimentation to values established by other methods suggests there is a technique error in neglecting multimode effects in cascading which are important for the calibration of a standard $\chi^{(3)}$ relative to cascaded $\chi^{(2)}$ parameters (33,35,47).

Other less definite yet important effects such as profile changes due to nonlinear refractive index alteration in spatially nonuniform high power beams must be carefully considered. As example, the use of nonidentical liquids and optical paths prior to and in, say, EFISH cells and the usual quartz calibration cells could cause potentially inaccurate $\chi^{(3)}$ determinations. Obviously these types of considerations are important when precise experimentation to test fine models of molecular behavior are intended, but have not stood as obstacle to uncovering the important general trends in molecular nonlinearity enhancement.

New THG Methods For Molecular Liquid Characterization. An area which is essential for understanding general third-order nonlinear polarizability is characterization of the purely electronic contributions (48). Several methods have been employed for this purpose on relatively simple species, but a more general survey of more complex compounds has not been possible due to the lack of a method of sufficient reliability and simplicity (48). Besides academic curiosity this area should be of some interest since there have been predictions of very large enhancements of electronic third-order hyperpolarizability, γ^e, (9) which could provide third-order susceptibilties capable of matching typical second-order responses at achievable electric field strengths. Since constraints on alignment of molecules to achieve substantial $\chi^{(3)}$ are very forgiving relative to the situation of $\chi^{(2)}$ described in the first part of this paper, there is strong technological motivation for work in this area. Another reason for interest in this possibility is that (nonexcitational) electronic responses are extremely fast unlike diffusion, rotation, or thermal contributions which dominate $\chi^{(3)}$ in most strong Kerr, nonlinear refraction, etc. media. For use in extremely high bandwidth applications, such as totally optical processing, predominantly electronic responding media are needed (49).

We have developed techniques for correctly (at least in relation to the issues raised in the last section) measuring $\chi^{(3)}$ in liquids by optical third harmonic generation (35,50). The $\chi^{(3)}$ of THG is most straightforward conceptually, suffering from the least complex dispersion and resonance contributions in the set of possible third-order susceptibilities (35,48). Because of the occurrence of only positive (summation of) frequencies, it may generally be assumed to be a measure of electronic nonlinearity alone. Furthermore, with the use of a sufficiently long wavelength fundamental, electronic dispersion enhancement can be minimized. This allows a more straightforward approach to the interpretation of relative nonlinearities, being less dependent on specific properties of the lowest few electronic excitation levels.

In two THG methods for characterization of liquids that we've devised, the interferences of air are eliminated by working in vacuum. Because weak focusing is employed (~1 m in our apparatus) the beam profile near focus can be considered nearly constant. The vacuum cell, of course, is also required to be long enough to avoid significant generation of $\underline{P}^{3\omega}$ at its windows. The obvious generalization of the wedge method is that the liquid chamber of the cell being placed in the apparatus be wedged. Unfortunately, since the windows of the cell also generate nonzero $\underline{P}^{3\omega}$ and since their phased contributions to the total $\underline{E}^{3\omega}$ must be known, there is little choice but to make them wedges as well (considering the prior discussion of the critical nature of ψ in condensed phase materials). We have therefore constructed such a cell in which two wedged windows of fused silica (f.s.) define a wedged liquid chamber. The shape of THG fringes generated when the cell is placed in vacuum and translated transverse to the optical path is complex but may be analyzed to extract ($\chi^{(3)}_{liq.}/\chi^{(3)}_{f.s.}$). In fact, this analysis of shape (not magnitude of signal relative to a secondary standard as in EFISH) is overdetermined and two independent evaluations are possible from the same data, giving an indication of reliability. This analysis is detailed and the reader is refered elsewhere for fuller exposition (35).

A second method was devised to simplify the complexity and laboriousness of the general three-wedge method (50). Here the cell is constructed with a long liquid chamber. If the liquid absorbs the laser weakly but nearly completely, $\underline{P}^{3\omega}\approx 0$ at the back window with significant simplification resulting. Since nearly all organic liquids have some weakly allowed high vibrational state transitions around 1.91 μm, this situation can usually be achieved with ~6 cm long liquid chamber for the 1.91 μm→636 nm THG which we have studied. Disregarding focusing, vibration diagrams describing THG in the absorbing liquids are spirals. Figure 6 shows the result of splicing a spiral onto the circle of the front window wedge. Here again the dashed curve represents the set of all possible $\xi=+\infty$ points due to the arbitrary value of ψ appropriate to the front window thickness. Simple intensity fringes are predicted with period of the window wedge, but with nonunity contrast ratio. Similar diagrams have been analyzed and it was found that evaluation both of the magnitude and of the contrast ratio of THG fringe patterns from such a liquid filled cell relative to a simple glass wedge allows two independent determinations of the ratio of liquid and window $\chi^{(3)}$'s. One might be somewhat concerned that the deposition of the entire energy of the laser pulse into the liquid during the short 20 nanosecond pulse duration might cause an unacceptable temperature rise and density variation. However, since the method relies on the establishment of a liquid contribution to $\underline{E}^{3\omega}$ which is proportional to $\chi^{(3)}/[\chi^{(1)}_{3\omega}-\chi^{(1)}_{\omega}]$, a much more weakly density dependent quantity than $\chi^{(3)}$, the impact of that effect is reduced.

Since only ratios of liquid to window susceptibilities were determined by these methods we have determined the ratios of $\chi^{(3)}$ in f.s., BK-7 and α-quartz (47) and have performed a cascading experiment to determine α-quartz $\chi^{(3)}$ relative to its $\chi^{(2)}$ (33). More detailed analysis of methods and results are available in the papers cited.

Further considerations of the goal to measure $\chi^{(3)}$ and avoid interferences has lead us to a more straightforward scheme than the

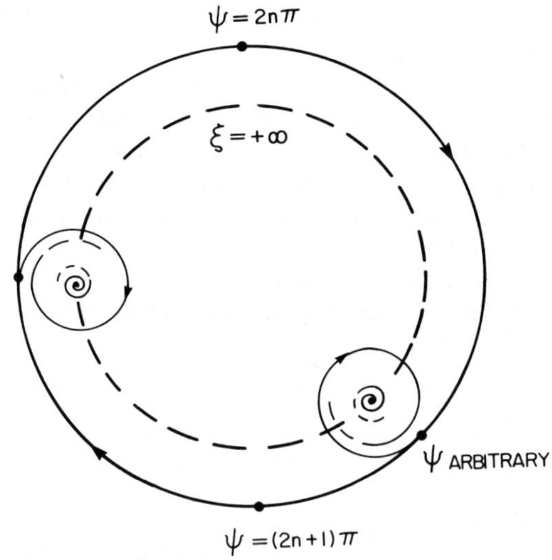

Figure 6. Vibration diagram depicting optical third harmonic generation when an absorbing liquid follows a thin solid. (See *text for further discussion.)*

two described above. Given that the THG $\chi^{(3)}$ has been measured for a number of liquids, a method which relies on relative intensities of THG might be used. We have devised an arrangement in which to achieve an accuracy of a few percent one needn't measure coherence lengths, just the relative intensities coupled with a refractive index measurement. Currently we are attempting realization of this method. It is anticipated that this same concept can be applied to the determination of EFISH nonlinearities.

Molecular Interpretation of Third-Order Susceptibilities. As stated in the introduction, development of new organic nonlinear optical media has relied on knowledge of molecular nonlinear polarizability. Experimentally, though, it is properties of $\chi^{(3)}$ which are determined. Molecular properties must be extracted by application of some model for their collective behavior in establishing $\chi^{(3)}$ (51). Description of $\chi^{(3)}$ in molecular liquids is a many-body problem, implying that most molecular interpretations are fundamentally inaccurate (48). An example of such an approximation is that for THG and EFISH the molecules are generally considered independent except through their mutual polarizing action. Even in this case one must somehow self-consistently describe the linear and nonlinear polarizations and microscopic fields experienced by each molecule of a statistical ensemble. In the dipole approximation this task can be achieved formally (50,51):

$$\chi^{(3)} = V^{-1}\,\Sigma_\lambda (\underline{N}^\lambda)^T \cdot \underline{\gamma}^\lambda \cdots \underline{N}^\lambda \underline{N}^\lambda \underline{N}^\lambda + \chi^{(3)\prime}$$

$$\chi^{(3)\prime} = V^{-1}\,\Sigma_\lambda (\underline{N}^\lambda)^T \cdot \underline{\beta}^\lambda \cdot\cdot \underline{N}^\lambda \Sigma_\mu (\underline{R}^{\lambda\mu})^T \cdot \Sigma_\nu \underline{L}^{\mu\nu} \cdot \underline{\beta}^\nu \cdot\cdot \underline{N}^\nu \underline{N}^\nu$$

where $\underline{N}^\lambda$ is a tensor which functionally relates the macroscopic and local electric fields at molecule λ, and $\underline{R}^{\lambda\mu}$ and $\underline{L}^{\lambda\mu}$ are tensors of the formalism which are associated with fields at molecule λ due to polarization of molecule μ. To be useful the results must be averaged over possible configurations of the liquid. The averaging processes involve many unknown factors.

For the moment disregarding $\chi^{(3)\prime}$, one is faced with the task of averaging $\Sigma_\lambda (\underline{N}^\lambda)^T \cdot \underline{\gamma}^\lambda \cdots \underline{N}^\lambda \underline{N}^\lambda \underline{N}^\lambda$. Steps which are necessary to proceed from this term to the commonly employed models employing averaged linear local field factors are:

1) Tensorial correlations among $\underline{\gamma}^\lambda$, $\underline{\alpha}^\lambda$ and the polarizability of the surrounding molecular ensemble which determine the details of $\underline{N}^\lambda$ are neglected; $\underline{\gamma}^\lambda$ is averaged independently of the other factors.

2) The tensorial product of local field tensors is not averaged, but replaced by an arithmetic product of the average magnitude of four local field factors.
3) Correlations between molecular orientations, that is, between terms with different λ, are disregarded (except in EFISH where the dipole correlation factor g is employed in the $\mu \cdot \beta^V$ term (40) - a practice which assumes the tensorial behavior of the vector components of $\underline{\beta}$ to show the same intermolecular correlations as the permanent dipole); single terms for each molecular type are adopted.
4) Frequently correlations of local field amplitude and molecular type are ignored and a single family of solution local field factors is adopted.

The two common local field models which are employed are the Debye-Lorentz and the Onsager models (1,40,52). Both replace the molecular solvent surroundings by a continuum with a cavity in which the λ molecule resides. Assuming a spherical cavity shape does not make the local field parallel to the macroscopic field due to the anisotropic polarizability (and orientation of the permanent dipole). At optical frequencies (details of the zero frequency local fields are beyond the scope of our investigations) a scalar averaged effective polarizability is assumed, establishing the parallelness. In the Debye-Lorentz model one assumes the local fields to be identical for all molecules. The Onsager model allows this average effective polarizability and the cavity size to be different for each molecular type. Simple estimations with parameters for a dilute solution of dissimilar molecules show that the product of solute optical local field factors (four for THG and three for EFISH) which multiply γ in this model of $\chi^{(3)}$ can differ by tens of percent between the two models. This means that solute γ's determined from $\chi^{(3)}$'s are subject to at least this magnitude uncertainty from the optical local fields! (This contradicts the conclusion of Singer and Garito (40), due to an unfortunate, simple algebraic error in the derivation of their expressions contrasting the two optical local field models.) Because of the importance of choosing the better model we have investigated the behavior of THG $\chi^{(3)}$ over the ranges of miscibility of several liquid mixtures (53). These studies also were intended to investigate the severity of the approximations 1)-4) above. However, although one might expect the high power of the local field factors to cause this to be a good test, the small contributions to $\chi^{(3)}$ of the highly deviating species at low concentrations and the play-off of the one increased contribution and the one decreased contribution in the Onsager model in the intermediate concentration region allowed both models to fit our data comparably well. We have seen by these studies, though, that beyond this uncertainty $\chi^{(3)}$ of THG is very well behaved compared to the intricacies associated with EFISH $\chi^{(3)}$ (40). In fact, simple analysis of the $\chi^{(3)}$ values of 21 neat liquids determined by the two THG techniques described above using these local field models (they are equivalent in neat liquids) shows remarkably simple apparent orientational averaged hyperpolarizability behavior (35,50). Systematics indicating validity of bond additivity are present. Even simple single ring aromatics are seen to be predictable to ~10% by that approach. The variation of accompanying "exaltations" of nonlinear polarizability of those compounds displays two phenomena. First, strong perturbation of the ring potential, as in pyridine or fluorobenzene, reduces the nonlinearity. Second, substitution with electron donors enhances nonlinearity while electron withdrawing groups diminish it. Analysis of results in a larger set of compounds is underway. A disappointing result which stood out in the set was the failure to observe the extremely large magnitude of γ^e predicted to occur in TCNQ, one of the molecules calculated (9) to have enhanced γ^e (54). In this case the solubility of TCNQ in common solvents was so low that an absolute value was not determined, but an upper limit on the magnitude of the orientational average of γ^e was established to be significantly below the expectation.

Returning again to the relationship between molecular and bulk properties, the $\chi^{(3)\prime}$ term above displays the important fact in that in condensed phase media nonlinearities beyond second-order are not purely and directly related to the corresponding molecular or microscopic nonlinearity (50,51,55,56). This is different than the cascading problem mentioned above in which the overall high-order response of a material is related not only to the corresponding high-order susceptibility, but also to sequential nonlinear interactions of lower-order involving the establishment of intermediate macroscopic electric fields. The suggestion here is that cascading of lower-order nonlinearity through local electric fields must be considered. The argument is straightforward. While the overall $\chi^{(2)}$ in a liquid vanishes and while each of the averaged local fields at a molecule due to the nonlinear response of other molecules of the liquid vanishes, there is a specific correlation in the summations of $\chi^{(3)\prime}$ which does not. The term with $\nu=\lambda$ describes the effect where molecule λ responds via β, that nonlinear polarization polarizes its surroundings, and the accompanying reaction field at molecule λ participates in another second-order polarization. This term does not average to zero because the β tensor of the same molecule enters twice in the process description. Details of this phenomena may be found in recently published works (50,55,56). As was mentioned in the first part of this paper, this analysis of THG in solutions of DANS and of the trans-4′-dimethylamino-N-methyl-4-stilbazolium ion in terms of $\chi^{(3)}$, has allowed us to conclude that the stilbazolium ion is nearly as nonlinear as DANS in second-order. (57) Obviously the corresponding EFISH measurement would have been difficult to acomplish.

Conclusion

We have briefly described several types of materials development research intended to generate either new highly nonlinear optical materials or new nonlinear optical materials having novel fabrication potential. This work utilized the inherent versatility and special properties of molecular materials. These and other novel approaches we are pursuing, and the several innovative approaches described in other papers of this volume are quite different than the practices of materials research in inorganic nonlinear media. While it is unrealistic to expect organics to displace the currently used inorganic nonlinear optical media, as this field progresses it is anticipated that applications employing special enabling characteristics of the organics will materialize.

Several aspects of third-order nonlinear optical experimentation and materials characterization associated with the development of organic nonlinear optical media were described in a largely nonmathematical manner. Pictorial models were employed to show the origin of several important nonintuitively obvious aspects of propagation. This opportunity was taken to point out several experimental pitfalls and describe errors of technique which have previously appeared in the literature. These models and the aspects they made obvious were used to devise new, more correct methods for third-harmonic generation characterization of the purely electronic third-order nonlinear polarization response of materials. Finally a brief discussion was given of limitations associated with describing $\chi^{(3)}$ in liquids in a molecular basis or of extracting molecular parameters from an experimentally determined liquid $\chi^{(3)}$.

Acknowledgments The author would like to acknowledge the collaborations of Drs. David Williams, Ronald Ziolo and Valeri Krongauz (visiting scientist, Weizmann Institute of Science, Israel) in addition to the contributions made by John VanDusen and Ronald Weagley, all employees at Xerox Webster Research Center. Finally the role of two students in this effort, Cheryl Hanzlik (APS Summer Intern, Univ. of Rochester) and Bonnie Buchalter (Cooperative Education Employee, Manhattan College) is acknowledged.

Literature Cited

1. Levine, B. F.; Bethea, C. G. J. Chem. Phys. 1975, 63, 266.
2. Oudar, J. L. J. Chem. Phys. 1977, 67, 446.
3. Ducuing, J. "Nonlinear Spectroscopy"; Bloembergen, N., Ed.; North-Holland: New York, 1977, p. 276.
4. Dulcic, A.; Fltyzanis, C. Opt. Comm. 1978, 25,402.
5. Jerphagnon, J.; Chemlas, D.; Bonneville, R. Adv. Phys. 1978, 27, 609.
6. Levine, B. F.; Bethea, C.G.; Thurmond, C. D.; Lynch, R. T.; Bernstein, J. L. J. Appl. Phys. 1979, 50, 2523.
7. Oudar, J. L.; Hierle, R. J. Appl. Phys. 1977, 48, 2699.
8. Zyss, J. S.; Chemla, D. S.; Nicond, J. F. J. Chem. Phys. 1981, 74, 4800.
9. McIntyre, E. F. ; Hameka, H. F. J. Chem. Phys. 1978, 70, 2215.
10. Kurtz, S. K.;Perry, T. T. J. Appl. Phys. 1968, 39, 3798.
11. Halbout, J. M.; Blit, S.; Tang, C. L. IEEE J. Quantum Electron. 1981, 17, 513.
12. Conwell, E. M. IEEE J. Quantm Electron. 1973, 9, 867.
13. Babai, F. H.; White, E. A. D. J. Crys. Growth 1980, 49, 245.
14. Morrell, J. A.; Albrecht, A. C.; Levin, K.; Tang, C. L. J. Chem. Phys. 1979, 71, 5063.
15. Lipscomb, G. F.; Garito, A. F.; Narang, R. S. J. Chem. Phys. 1981, 75, 1509.
16. Sigelle, M.; Hierle, R. J. Appl. Phys. 1981, 52, 4199.
17. Byer, R. L. "Nonlinear Optics"; Harper, P. G.; Wherrett, B. S., Eds.; Academic: New York, 1977, p.47.
18. Zernike, F.; Midwinter, J. E. "Applied Nonlinear Optics"; Wiley: New York, 1973.
19. Ziolo, R. F.; Gunther, W. H. H.; Meredith, G. R.; Williams, D. J. Acta Crys. 1982, B38, 341.
20. Kitaigorodsky, A. I.; "Molecular Crystals and Molecules"; Academic: New York, 1973.
21. Lu, T. H.; Lee, T. J.; Wong, C.; Kuo, K. T. J. Chin. Chem. Soc. 1979, 26, 53.
22. Levine, B. F.; Bethea, C. G.; Wasserman, E.;Leenders, L. J. Chem. Phys. 1978, 68, 5042.
23. Meredith, G. R.; Ziolo, R. F.; Williams, D. J.; Weagley, R. J. Am. Chem. Soc. submitted.
24. Krongauz, V. A. Israel J. Chem. 1979, 18, 304.
25. Meredith, G. R.; Krongauz, V.; Williams, D. J. Chem. Phys. Lett. 1982, 87, 289.
26. Meredith, G. R.; Williams, D. J.; Fishman, S. N.; Goldburt, E. S.;Krongauz, V. J. Phys. Chem. to appear.
27. Havinga, E. E.; VanPelt, P. Ber. Bunsenges. Phys. Chem. 1979, 83, 816.

28. Meredith, G. R.; VanDusen, J.; Williams, D. J. Macromolecules 1982, 15, 1385.
29. Meredith, G. R. Rev. Sci. Instrum. 1982, 53, 48.
30. Bloembergen, N. "Nonlinear Optics"; Benjamin: Reading, 1965.
31. Jackson, J. D. "Classical Electronics"; Wiley: New York, 1975.
32. Ward, J. F.; New, G. H. C. Phys. Rev. 1969, 185, 57.
33. Meredith, G. R. Phys. Rev. 1981, B15, 5522.
34. Meredith, G. R. Opt. Comm. 1981, 39, 89.
35. Meredith, G. R.; Buchalter, B.; C. Hanzlik J. Chem. Phys. 1983, 78, 1543.
36. Kleinman, D. A.; Miller, R. C. Phys. Rev. 1966, 148, 302.
37. Kurtz, S. K. "Quantum Electronics: A Treatise"; Rabin, H.; Tang, C. L., Eds.; Academic: New York, 1975, Vol I., p. 209.
38. Bethea, C. G. Appl. Opt. 1975, 14, 2435.
39. Garito, A. F., personal communication
40. Singer, K. D.; Garito, A. F. J. Chem. Phys. 1981, 75, 3572.
41. Flytzanis, C.; Bloembergen, N. Prog. Quantum Electron. 1976, 4, 271.
42. Hermann, J. P. Opt. Comm. 1973, 9, 74.
43. Meredith, G. R. J. Chem. Phys. 1982, 77, 5863.
44. Hochstrasser, R. M.; Meredith, G. R.; Trommsdorff, H. P. Chem. Phys. Lett. 1978, 53, 423.
45. Levenson, M. D.; Bloembergen, N. Phys. Rev. 1974, B10, 4447
46. Song, J. J.; Levenson, M. D. J. Appl. Phys. 1977, 48, 3496.
47. Buchalter, B.; Meredith, G. R. Appl. Opt. 1982, 21, 3221.
48. Hellwarth, R. W. Prog. Quantum Electron., 1977, 5, 1.
49. Smith, P. W.; Tomlinson, W. J. IEEE Spectrum 1981, 18(6), 26.
50. Meredith, G. R.; Buchalter, B.; C. Hanzlik J. Chem. Phys. 1983, 78, 1533.
51. Flytzanis, C. "Quantum Electronics: A Treatise"; Rabin, H.; Tang, C. L., Eds.; Academic: New York, 1975, Vol I., p. 9.
52. Bottcher, C. J. F "Theory of Electric Polarization"; Elsevier: New York, 1973.
53. Meredith, G. R.; Buchalter, B., unpublished data.
54. Meredith, G. R.; Buchalter, B. J. Chem. Phys. 1983, 78, 1615.
55. Meredith, G. R. Chem. Phys. Lett. 1982, 37, 516.
56. Meredith, G. R.; Buchalter, B. J. Chem. Phys. 1983, 78, 1938.
57. Meredith, G. R., unpublished data.

RECEIVED July 29, 1983

3

Organic Materials for Optical Second Harmonic Generation

ROBERT J. TWIEG and KANTI JAIN

IBM Research Laboratory, San Jose, CA 95193

Recent studies have revealed the anomalously large optical nonlinearities of certain organic compounds relative to those of more conventional inorganic substances. Progress in the optimization of the optical nonlinearity of organic materials has stemmed from the interactive combination of better theoretical understanding of the origins of these large nonlinearities and the rational modification of molecular and crystal structure via organic synthesis.

Just as great strides have been made in electronics and semiconductors, so will optics evolve in the future to solve a variety of technological problems. Both optics and electronics, which have traditionally relied upon inorganic materials for fabrication of various components, will benefit greatly from the plethora of organic compounds, millions already known and many more awaiting preparation and application. Already the impact of organic compounds has been felt, to specify a few instances: organic dyes for lasers, holographic media and photoconductors; liquid crystals in displays and optical switches; piezoelectric polymers; conducting and superconducting organic solids; and the multitude of organic photoresists and polymers involved in the fabrication of integrated circuits.

Of the many potential applications of organic materials to optics technology we will focus on but the single facet of frequency doubling via second harmonic generation (SHG) in which light with a fundamental frequency ω interacts with a material in such a fashion that conversion to 2ω occurs. This phenomenon is already widely used for doubling fundamental wavelengths from the IR into the UV, for producing radiation of wavelength suitable for pumping dyes, and for the analysis of short pulses. Our own work to date has concentrated on crystalline materials potentially useful for doubling the output of semiconductor lasers with a fundamental wavelength of

0097–6156/83/0233–0057$07.00/0

0.80-1.50μ, in particular those in the GaAs family with fundamental outputs at wavelengths less than 1.0μ. A device based on the frequency doubled GaAs laser will find a variety of practical applications in electrophotography, scanning, optical storage, and a variety of other important technologies. For a given organic material the most important properties which must be considered for such applications are doubling efficiency, transparency, phase matching characteristics and crystal growth parameters. Although most of the discussion which follows concentrates on efficiency and transparency, all these properties are ultimately of equal importance for the practical implementation of the SHG phenomenon in a working device.

It has been just twenty years since the SHG phenomenon manifested itself in the feeble activity of quartz (1). This result was made possible by the just prior discovery of the laser which for the first time provided the intense monochromatic radiation required to observe the nonlinear effect. Shortly thereafter a complete theoretical treatment of the SHG phenomenon in nonlinear media appeared (2). Subsequent important developments included the powder technique of Kurtz and Perry (3) which allowed the rapid assay of materials for which high quality crystals were not available, the demonstration of the importance of phase matching (4) critical for efficient SHG to occur, and later, the ability to test for phase matching on powder samples (5). In their experiments on the powder technique Kurtz and Perry also demonstrated SHG from a variety of organic compounds which were selected on the basis of their noncentrosymmetric crystal structure. The first tests of organic compounds were highly empirical in nature (and, alas, remain so even today!); however, important relationships between the structure of a compound and its SHG activity have been derived and supported on theoretical grounds. Already crystalline compounds two orders of magnitude more efficient than urea (commonly employed as a reference) are known and even higher efficiencies have been found for certain dyes in electric field induced experiments. It is not unreasonable to expect that significant progress will continue to occur in the near future.

We will begin our discussion with a short description of the physical origin of nonlinear effects such as SHG, present some of the typical organic nonlinear materials which have appeared in the literature and then examine some of the important theoretical developments which attempt to relate SHG activity with molecular and crystal structure. Due to the brevity of this chapter it is impossible to explore these in any great depth and so the interested reader is directed to two excellent short reviews on organic materials for SHG (6,7), to two relevant reviews on the organic solid state (8,9), to the other chapters in this book (especially for SHG in solution and polymers) and to the leading references herein. Finally, we will present work on materials discovered in our own laboratory and the rationale which went into their design and preparation.

Origin of the SHG Phenomenon

The fundamental equation (1) describes the change in dipole moment between the ground state μ_g and an excited state μ_e expressed as a power series of the electric field E which occurs upon interaction of such a field, as in the electric component of electromagnetic radiation, with a single molecule. The coefficient α is the familiar linear polarizability, β and γ are the quadratic and cubic hyperpolarizabilities, respectively. The coefficients for these hyperpolarizabilities are tensor quantities and therefore highly symmetry dependent; odd order coefficients are nonvanishing for all molecules but even order coefficients such as β (responsible for SHG) are zero for centrosymmetric molecules. Equation (2) is identical with (1) except that it describes a macroscopic polarization, such as that arising from an array of molecules in a crystal (10).

$$\Delta\mu = \mu_e - \mu_g = \alpha E + \beta EE + \gamma EEE + \ldots \quad (1)$$

$$P = P_0 + \chi^{(1)}E + \chi^{(2)}EE + \chi^{(3)}EEE + \ldots \quad (2)$$

Light waves passing through an array of molecules can interact with them to produce new waves; this interaction may be interpreted as resulting from a modulation in refractive index or alternatively as a nonlinearity of the polarization. Such interaction occurs most efficiently when certain phase matching conditions are met, requiring identical propagation speeds of the fundamental wave and the harmonic wave. Birefringent crystals often possess propagation directions in which the refractive index for the fundamental ω and the second harmonic 2ω are identical so that dispersion may be overcome.

Typical Organic Nonlinear Materials

Table I lists a variety of organic nonlinear materials which have appeared in the literature; their relative powder efficiencies, absorption cutoffs and β values (if available) are also provided. These materials are "typical" only in that they represent results from the few classes of organic compounds investigated to date, yet they are instructive in that one learns which molecular properties may be important. A few caveats are in order to avoid misinterpretation of the data in Table I. Except for compound 10 (19) all the powder efficiency and cutoff data are from our own measurements. Powder measurements were performed on ungraded samples using the Nd:YAG output at 1.06μ as fundamental; since powder efficiency is a function of particle size distribution and a variety of other factors (3) these values are only semiquantitative. The cutoff values are the wavelengths for which $\sim 10^{-4}$M solutions in ethanol (unless otherwise indicated) have no absorbance. The cutoff values will be similar to those found in crystal state except where intermolecular charge transfer is important in the crystal or the molecule is solvatochromic, this latter effect being quite common for cyanine dyes such as

Table I. Typical Organic Nonlinear Materials

#	Structure	Eff (x urea)	cutoff (nm)	β ($\times 10^{-30}$ esu)	Ref.
1	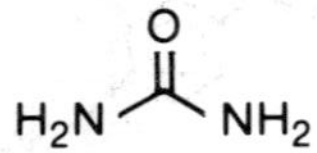	1.0 [x2.5]	200	0.45	3,11
2	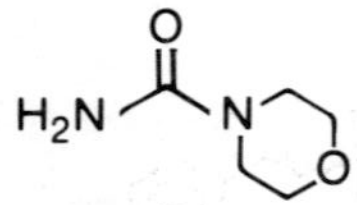	0.0	~210	—	12
3	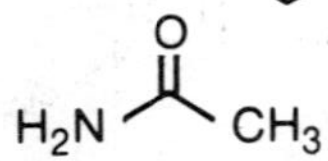	0.3	~210	—	3,5
4	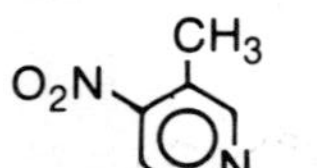 POM	13 [x4]	410	5	13
5	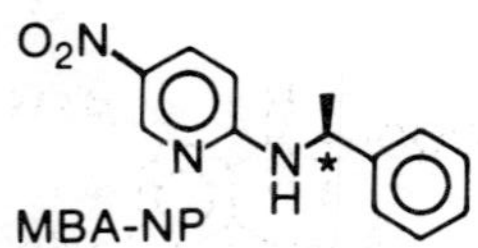 MBA-NP	25	430	—	14
6	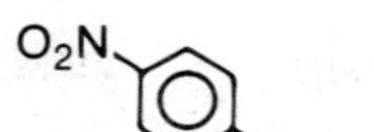 NA	0.0	470	35	15,16

7	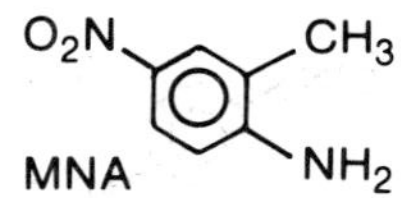	22 [x3.5]	480	42	15,16
8	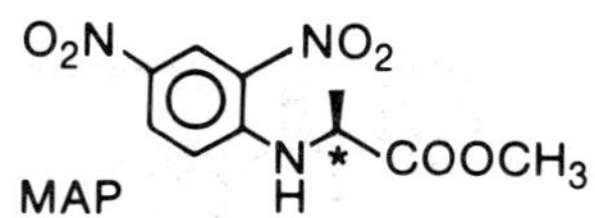	10 [x6.7]	500	220	17
9	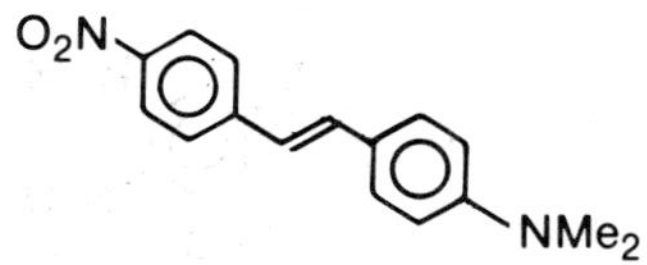	0.0	430 (λ_{max}) ~580	450	18
10	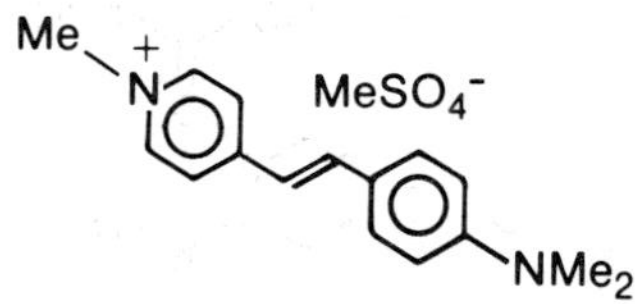	~250	473 (λ_{max}) 588 (MeOH)	—	19
11	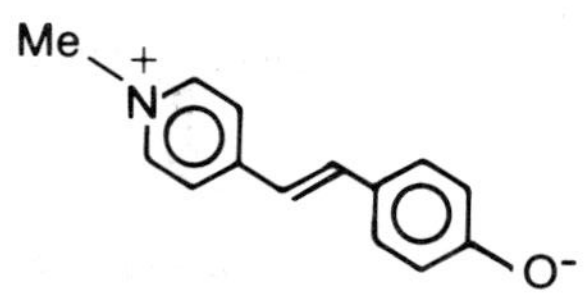	0.0	570 (λ_{max}) ~650	1000	20

[the bracketed efficiency factors apply to crystallographically optimal materials, see text]

10 and 11. Likewise, the β values determined in solution will be dependent on the solvent and a wide variety of other experimental parameters. The purpose of Table I is not to compile precise absolute data on these compounds but simply to demonstrate certain trends between structure and optical nonlinearity.

After perusal of Table I one can make two generalizations; first, active molecules are conjugated and second, they are polarized, that is, they contain donor and acceptor (D-A) substituents. The presence of a polarized conjugated system is not mandatory for SHG activity but these two features are often found in those materials which are the most efficient. Close examination of the compounds in Table I reveals some of the pitfalls inherent in making generalizations about structure and SHG activity. Urea 1, 4-morpholinecarboxamide 2 and acetamide 3 all contain the minimal amount of conjugation and polarization in the carbonyl group but activity in the crystalline state is seen to be dependent on the specific molecule and its substituents. Urea and acetamide are both well known to crystallize in a noncentrosymmetric fashion and so β is nonvanishing but 4-morpholinecarboxamide has zero efficiency, likely due to centrosymmetric crystallization. Although the powder efficiency and β value of urea are small compared to most of the organic molecules we will encounter, one must recognize that urea has nonlinear properties comparable to the best inorganic materials. The only serious shortcoming of urea relative to many inorganic frequency doublers is the difficulty encountered in growing large high quality crystals. However, constant progress is being made in the area of organic crystal growth and as experience accumulates this difficulty will be gradually overcome (21).

Compounds 4 through 9 represent some of the most extensively studied materials in terms of theoretical treatment, crystal growth, phase matching studies, and the like. All are polarized aromatic molecules, yet here again SHG activity is highly dependent on the specific structure; consider, for example, 4-nitroaniline 6 (NA) and 2-methyl-4-nitroaniline 7 (MNA). The β values for these two molecules are similar but the SHG powder efficiency of NA is zero due to centrosymmetric crystallization whereas MNA, differing only by the presence of a methyl group, crystallizes noncentrosymmetrically and thus the intrinsic nonlinearity of the nitroaniline system can manifest itself in the crystalline state. The same is true of the more highly conjugated and polarized styrylpyridinium cyanine dye 10 and merocyanine 11 The β values of these cyanine dyes are probably comparable (the β of 10 cannot be determined without independent determination of its dipole moment) but only 10 crystallizes noncentrosymmetrically and is active in the crystalline state. It is clear from the table that (going from top to bottom) increased conjugation and polarization leads to larger β values and a corresponding shift in absorption edge from the UV into the visible. The powder efficiency, however, cannot be correlated as well with β because of intervening contributions due to crystallization, which is so highly dependent on the specific molecular structure. Even those crystalline compounds which are active usually do not take full advantage of the intrinsic nonlinearity of the molecules from which they are composed. The numbers in brackets in Table I give the enhancements in powder efficiency one might anticipate if the molecules were optimally aligned in the crystal (more accurately these values apply to enhancements of d values) (22). We will see many more examples of apparently subtle structural changes producing enormous differences in the crystal state SHG activity of organic compounds. The design of such molecules requires a delicate balance of a variety of parameters, some so mutually interactive that they defy separation, rendering the undertaking all the more challenging.

Molecular Structure and SHG

Here we will treat molecular properties responsible for SHG excluding for the moment the additional constraints imposed by crystal state properties resulting from an array of such molecules. Fundamental to this discussion is the vector dipole moment of a molecule $\vec{\mu}$ given by Equation (3) as the product of the magnitude of charge separation e and the length of charge separation ℓ (23). In an aliphatic system, dipole moments are generally small but may be large as in the case of certain long chain α,ω-amino acids having fully charged terminal functionality in which conformationally dependent dipoles up to 30D have been found (24). Large dipole moments may also result from only partial charge transfer between donor and acceptor groups terminating a conjugated chain, as is found in certain merocyanine dyes where dipole moments in the range of 10-20D are not uncommon (25). A substituent mesomeric moment μ_m which results from the polarization of a π-electron system by a substituent is defined by Equation (4) as the difference between the dipole moment of an aromatic molecule μ_{ar} and an aliphatic molecule μ_{al} bearing that same substituent. Thus, the substituent mesomeric moment is a measure of the polarization of a π-electron system by a substituent. For cases involving two substituents on a π-electron system the interaction moment μ_{int}, a measure of isovalent mesomerism, is given by Equation (5) as the difference between the dipole moment μ_{ab} of the disubstituted molecule and its monosubstituted counterparts μ_a and μ_b. Equation (1) indicates that $\Delta\mu$ is the important quantity, not μ_g or μ_e alone. However, in general, molecules with large ground state dipole moments have correspondingly large excited state dipoles and so treatment of ground state dipoles will usually (but not always!) suffice.

$$\vec{\mu} = e\vec{\ell} \tag{3}$$

$$\mu_m = \mu_{ar} - \mu_{al} \tag{4}$$

$$\mu_{int} = \mu_{ab} - \mu_a - \mu_b \tag{5}$$

In discussing the origins of optical (hyper)polarizabilities in relation to molecular dipole moments it is convenient to separate the contributions due to the length of the conjugated system and the substituent perturbations on it. The linear polarizability is well known to be proportional to the third power of conjugation length, $\alpha \propto L^3$ (26), by the use of a free electron model it has been predicted that $\gamma \propto L^5$ (27), and by a combination of this relation and the EIF model it has been predicted that $\beta \propto L^3$ (12). The cubic hyperpolarizabilities for a "homologous" series of polyenes containing three to nineteen double bonds increase by a factor of 5×10^3 as a function of length in good agreement with theory. The cubic polarizabilities for comparable nonconjugated saturated molecules ranging from methane to pentadecane increase at a much slower rate and differ only by a factor of five (27). More qualitative relationships between conjugation length and the quadratic hyperpolarizability have been found; for example, the β values for substituted stilbenes are consistently larger by an order of magnitude than those for comparably substituted benzene analogs (18). Overall, the relationships between conjugation length and hyperpolarizability are well established both theoretically and by experiment.

Based on the fundamental dipole moment concepts of mesomeric moment and interaction moment, models to explain the enhanced optical nonlinearities of polarized conjugated molecules have been devised. The equivalent internal field (EIF) model of Oudar and Chemla relates the β of a molecule to an equivalent electric field E_R due to substituent R which biases the hyperpolarizabilities (28). In the case of donor-acceptor systems anomalously large nonlinearities result as a consequence of contributions from intramolecular charge-transfer interaction (related to μ_{int}) and expressions to quantify this contribution have been obtained (29). Related treatments dealing with this problem have appeared; one due to Levine and Bethea bearing directly on the EIF model (30), another due to Levine using spectroscopically derived substituent perturbations rather than dipole moment based data (31) and yet another more empirical treatment by Dulcic and Sauteret involving reinforcement of substituent effects (32).

It is of interest to note that almost fifty years ago physical organic chemists began to derive the corresponding relationships between substituents and chemical reactivity. These are the well known free energy relationships such as the Hammett Equation (6) in which σ is a substituent constant, ρ a reaction constant, k_R the rate of the reaction with substituent R present and k_o the rate of reaction with a standard substituent (usually hydrogen) (33).

$$\rho\sigma = \log(k_R/k_o) \tag{6}$$

A large number of modifications and refinements have been made on this equation, the most relevant being those which attempt to separate inductive σ_I and mesomeric σ_R contributions of a substituent as is found in the Taft equation. In addition to substituent constants based on reactivity, acidity and the like, a variety of spectroscopically derived constants such as those from fluorine NMR σ_R(shielding) (33,34) and IR band intensities R° (33,35) have more recently appeared in the literature. We will utilize some of these R° values to help explain the nonlinear properties of certain molecules in later discussions. The wealth of information obtained from these free energy relationships provides the organic chemist with qualitative insight into donor and acceptor substituents useful for the design and custom synthesis of optically nonlinear organic compounds.

The presence of conjugation and D-A groups introduces a generally undesirable side-effect, the so-called "transparency-efficiency tradeoff". The incremental increase in conjugation by linking double bonds produces a gradual bathochromic shift in absorption edge from 200 nm in ethylene to a maximum in the range of 600-700 nm for an infinite series of double bonds. Addition of D-A groups to a conjugated system brings about a further dramatic bathochromic shift in absorption edge as a result of stabilization due to mixing of nonbonded and charge-transfer states (36). An excellent example of this effect is found in a comparison of the cutoffs for the parent stilbene molecule (350 nm), D-A substituted 4-dimethylamino-4'-nitrostilbene 9 (580 nm) and the heterocyclic dye merocyanine 11 (650 nm) which all contain essentially the same size conjugated system. In terms of practical laser frequency doubling the absorption edge should be near the wavelength of the second harmonic but must not include it and so the material and the laser it is to double must be closely matched. The transparency-efficiency tradeoff is a nuisance, to say the least, but molecules such as POM, 4, have been designed in such a way to minimize the ground state dipole and thereby simultaneously favorably effect absorption edge and crystal packing as well (13).

Crystal Structure and SHG

We have just described some optimal molecular parameters for organic compounds intended for use in SHG. There exists a set of comparable parameters which arise from dealing with an array of molecules, as in a crystal. In general, these parameters are more subtle, less well understood and ultimately more difficult to control. In constructing molecules the chemist deals with covalent bond energies generally of 50 to 200 kcal/mol; a typical molecule is made up of dozens of such bonds and the total bonding energy of the molecule is on the order of thousands of kcal/mol. In contrast, the energies involved in binding molecules to one another in crystals are significantly less. The crystallization of molecules is influenced by molecular shape (the close-packing principle), van der Waal's interactions (10-20 kcal/mol for an "average" size molecule), hydrogen bonding (if present, 3-6 kcal/mol per bond) and multipolar interactions (<5 kcal/mol for an "average" molecule). The multipolar interaction energies are dependent on the crystal space group and, in the relevant case of dipolar interactions, proportional to μ_g^2. The dipole energy contribution E_μ to total sublimation energy (a measure of the sum of forces binding molecules together in the crystal) of a typical substance varies slowly with structural parameters and will not bring structures out of a position of minimum energy resulting from van der Waal's interaction in a close-packing situation (37,38). However, as we have seen, molecules designed for SHG are not typical in that their ground state dipoles are often large. In such a case the μ_g^2 interaction tends to favor crystal structures which are centrosymmetric so that dipoles may oppose one another or at the least bring the dipoles out of the required net alignment required for efficient SHG. The hydrogen bond energy is comparable to E_μ and it is becoming increasingly clear that hydrogen bonding can, in some instances, play an important role in "neutralizing" other forces and bringing about a favorable orientation of molecular dipoles.

Organic compounds can crystallize in any of 230 unique space groups which are divided into 32 classes of crystallographic point groups by symmetry. Of these most of the enantiomorphic classes (1, 2, 222, 4, 3, 32, 6, 23) and ten other classes (m, mm2, $\bar{4}$, $\bar{4}$2m, 4 mm, 3m, $\bar{6}$, 6 mm, $\bar{6}$m2, $\bar{4}$3m) are active for SHG. All the other classes are eliminated due to centrosymmetry or Kleinmann relations. Statistical surveys indicate that about 90% of organic compounds crystallize in a centrosymmetric class which leaves at most only 10% of all organic compounds which crystallize in a noncentrosymmetric class and therefore potentially useful for second harmonic generation. In addition, the surveys indicate that some 80% of the useful compounds will belong to either of two space groups, $P2_1$ or $P2_1P2_1P2_1$ (39). This severe statistical constraint is offset in part by the use of pure enantiomers which must crystallize noncentrosymmetrically (17,40). Although this guarantees some activity, one finds, in practice, great differences in efficiency between related optically active compounds as will become evident in the tables which follow. Also, one must be prepared to deal with the synthetic complications arising from the use of optical activity, including racemization, optical purity, resolution, and the like. To complicate matters even further there also exist certain symmetry class dependent dipole moment orientations which must be optimized to allow for efficient phase matching. These important relationships have only very recently been derived by Oudar and Zyss and should prove very useful for interpreting SHG efficiency (22).

Our Approach to the Design of Organic Molecules for SHG

When we began our work on organic frequency doubling materials the best known and studied compounds were urea and a few derivatives of nitroaniline. As a first attempt at custom design of organic molecules we felt it would be worthwhile to combine the favorable noncentrosymmetric crystal properties of urea with the large nonlinearity of nitroaniline. We screened a wide variety of urea and nitrobenzene derivatives and found that DNP-SC **12** and DNPU **13** were the most efficient (12,41). The structures of these materials are found in Table II along with two other diarylureas derived from two classes which have not been as thoroughly explored, examples of which are the pyridylurea **14** and the D-A substituted urea **15** derived from 2-fluoro-5-nitroaniline.

Table II. Urea Derivates

		Powder Eff. (x urea)
12	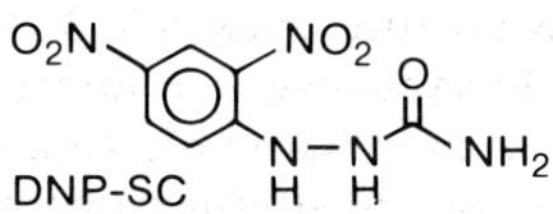 DNP-SC	8.8
13	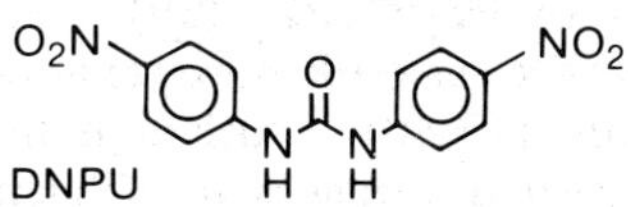 DNPU	8.8
14	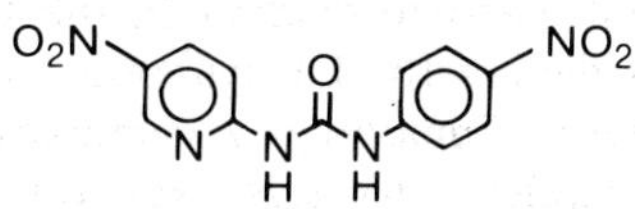	0
15	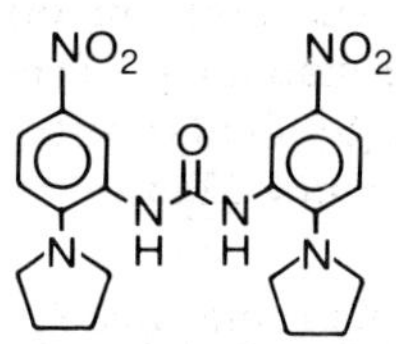	0

During this screening we also found the compound DMA-NAP 16, a highly polarized enone, had an efficiency some 7.5 times that of urea. This structure type is somewhat reminiscent of diethylaminomethylcoumarin which which had been studied previously in some detail (42). With these two precedents as guides we then prepared a series of other polarized enones, particularly the readily prepared chalcones and benzylidene acetones of which the compounds in Table III are typical. The most efficient derivatives all contained the nitrobenzene ring, although its presence is not essential for activity as in 4,4'-dimethylchalcone 19, which is well known to form an enantiomorphous crystal in class $P2_12_12_1$ (43). Remarkably, a comparable series of compounds (not depicted) in which the nitrile group replaced the nitro group were all devoid of activity.

The optically active amino acid derivative of dinitrobenzene, MAP, 8, is one of the most thoroughly studied organic nonlinear materials. A variety of related derivatives obtained by changing the amino acid functionality had already been prepared and examined (7,44). We prepared new analogs, compounds 22-32 listed in Table IV, in which both the amine donor and the substituents on the aromatic ring were changed. The range of activity observed for these electronically similar molecules indicates the important role that small changes in structure can play in determining crystal packing, and in turn, SHG efficiency. A wide variety of other dinitrobenzene derivatives of achiral amines were also prepared, but none showed any significant SHG activity.

The nitroaniline derivative MNA 7 is also a thoroughly studied organic nonlinear material (15,16). We decided to make a variety of MNA analogs from both racemic and optically active amines. Of these new materials, listed in Table V, the only achiral derivative with any activity is MNMA 33 in which the donor group is now methylamino instead of amino. Powder measurements indicate that MNMA is some four times as efficient as MNA, but the origins of this enhanced activity are somewhat difficult to delineate since MNMA crystallizes in the orthorombic space group Pna2-1 whereas MNA crystallizes in the monoclinic space group Cc. One component of this enhanced activity is certain, that being the better donor properties of the methylamino group versus the amino group (R°=-0.52 vs. -0.47) (33,35). Hydrogen bonding between the amino hydrogen and the nitro group of an adjacent molecule plays an important role in the crystal packing of MNMA and thereby contributes to the large SHG activity of this molecule. It is of interest to note that the optically active derivatives of MNA we have prepared are all significantly less efficient than either MNA or MNMA. Also, a number of analogous 2-trifluoromethyl and 3-methyl analogs of MNA (not depicted) showed little or no SHG activity.

A shortcoming of the DNPU molecule 13 is that the hydrogen bonding urea group desired for noncentrosymmetric crystallization is a rather poor donor even relative to amino (R°=~-0.40 vs. -0.47) (33,35). We decided to look at other nitroaniline derivatives with the hydrogen bonding group placed at alternative locations in the molecule. For this purpose the readily available 2-fluoro-5-nitroaniline serves very well as a synthetic starting material. The

Table III. Polarized Enones

16	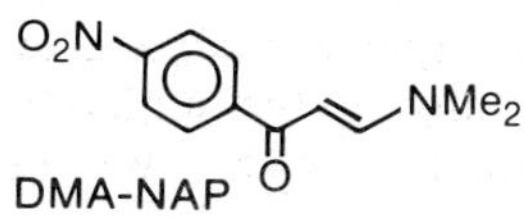 DMA-NAP	7.5
17	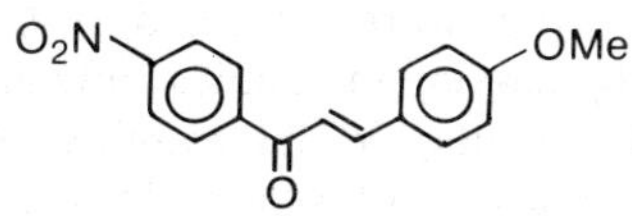	5.0
18	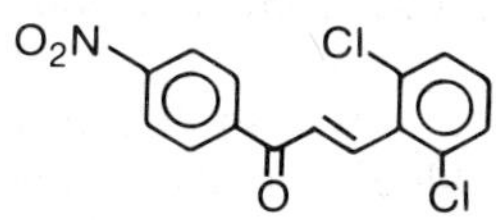	4.0
19	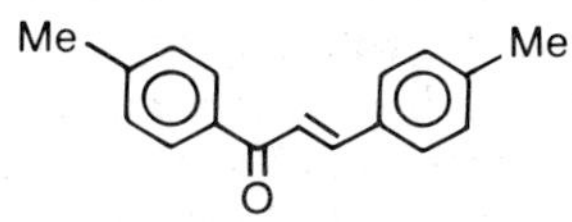	1.0
20	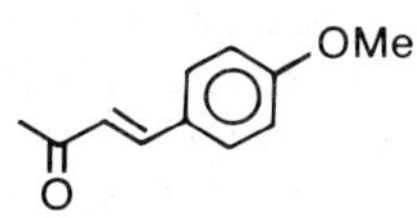	1.0
21	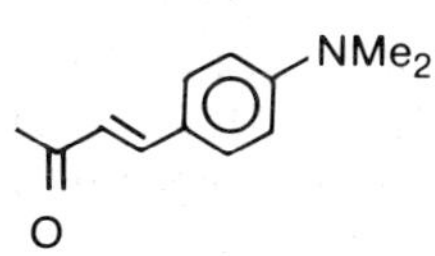	weak

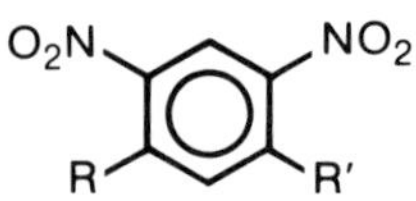

Table IV. SHG Efficiency of MAP Analogs

R	R'		
	H	F	Cl
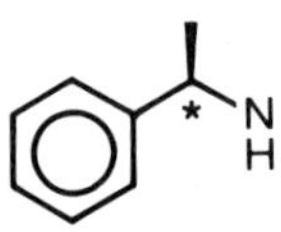	22 NC	23 1	24 <1
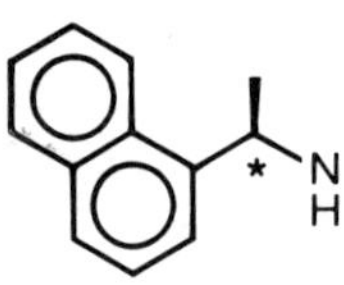	25 16	26 10	27 17
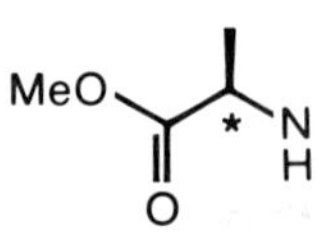	8 10 MAP	28 21	29 2
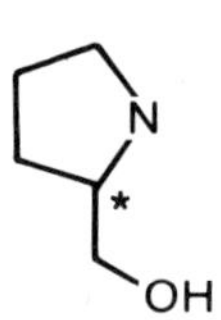	30 NC	31 3	32 9

NC = not crystalline

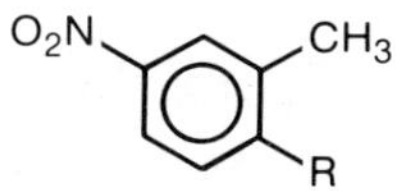

Table V. MNA Analogs

#	R	Powder Eff (x urea)
7	NH_2 (MNA)	22
33	HNMe (MNMA)	80
34	HNEt	0
35	$HNCH(CH_3)_2$	0
36	$HNCH_2CH_2OH$	0
37	$HNCH_2CH_2CH_3$	0
38	NMe_2	0
39	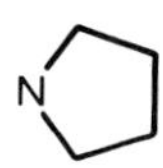	0
40	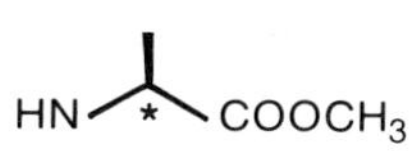	weak
41	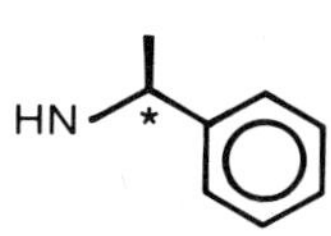	6
42	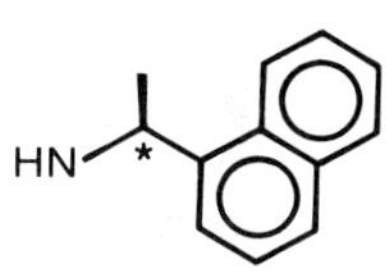	weak

aniline functionality is first converted to one of a variety of potentially hydrogen bonding groups such as acetamide, trifluoroacetamide, propionamide, formamide, ureido and succinimide and then the fluorine is replaced by a donor by nucleophilic aromatic substitution. Without such prior derivatization of the aniline the nucleophilic aromatic substitution reaction is very sluggish. Table VI lists representative members of this new class of nitroaniline based nonlinear materials. When R' is acetyl and R (the amine donor) is gradually changed, 40-45, dramatic changes in SHG efficiency occur. The most efficient compounds in this group are DAN 42 (115 x urea) and PAN 44 (80 x urea) which contain the dimethylamino and pyrrolidine donors, respectively. A crystal structure determination of DAN (space group P2-1) shows that hydrogen bonding occurs between the amide hydrogen in one molecule and the amide carbonyl of an adjacent molecule. This hydrogen bonding plays a significant role in aligning the molecular dipoles so that efficient SHG may result. The relatively low efficiency of 45 is typical of the results we have obtained with optically active donors in this class. When R' is propionyl as in compounds 46-48 the incidence and efficiency of SHG is greatly reduced and when R' is trifluoroacetyl the only compound found with any significant activity is 49 (70 x urea). Only optically active amines give active derivatives, such as 50, from the succinimide substituted compounds, and the same is true of the formamide and ureido systems prepared to date. Because we are able to simultaneously modify two functional groups in this class of compounds a great deal of flexibility for molecular engineering exists and we anticipate that more efficient compounds may be found (45).

Nonlinear effects are by no means limited to D-A substituted benzene derivatives, and, in fact, the bulk of our effort has been devoted to exploring heterocyclic substrates, especially derivatives of pyridine. We felt there were at least two reasons to examine pyridine based nonlinear materials: first, the cutoffs of the 2-substituted nitropyridines are consistently hypsochromically shifted about 30 nm from their benzene analogs (this consideration is especially important when we are trying to match organic doublers to a semiconductor laser) and, second, the halopyridines are generally more reactive than their benzene counterparts, which makes synthesis of analogs somewhat easier. During the initial screening of commercially available nitropyridine derivatives we found that 2-chloro-3,5-dinitropyridine 51 was some five times as nonlinear as urea in powder form. This discovery was important not due to the nonlinearity of this single substance, but moreover the recognition that it serves as a precursor for nitropyridine derivatives such as are listed in Table VII. A wide variety of 2-substituted-5-nitropyridines and 2-substituted-3-methyl-5-nitropyridines were also prepared and the best of these are also presented in Table VII. A number of these nitropyridines 51, 52, 53, 59 are of interest since they do not contain an amine donor yet possess modest SHG activity. Among the many nitropyridine derivatives we have examined the pure enantiomers prepared from optically active amines consistently gave the best results. In the case of racemic amines only two, 54 and 61, out of about a dozen such compounds were active. In addition, we

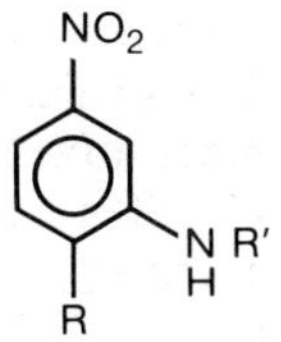

Table VI. Bifunctional Nitrobenzenes

#	R	R'	Powder Eff. (x urea)
40	NH_2	$COCH_3$	20
41	$HNCH_3$	$COCH_3$	weak
42	NMe_2	$COCH_3$ (DAN)	115
43	NMeEt	$COCH_3$	2
44	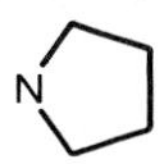	$COCH_3$ (PAN)	80
45	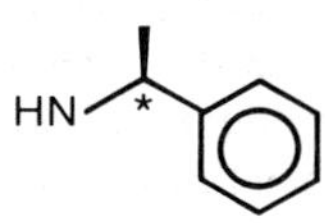	$COCH_3$	0.7
46	NH_2	$COCH_2CH_3$	10
47	NHMe	$COCH_2CH_3$	5
48	NHEt	$COCH_2CH_3$	0
49	NMe_2	$COCF_3$	70
50	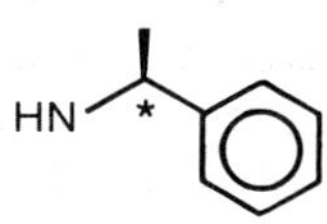	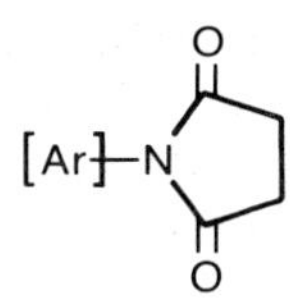	8

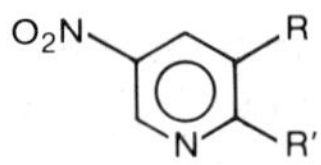

Table VII. Nonlinear Nitropyridines

#	R		R′		Powder Eff. (x urea)
51	NO_2		Cl		5
52	NO_2		OH		5
53	NO_2		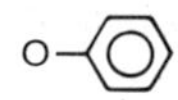		7
54	NO_2	(+,—)	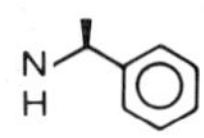		4
55	NO_2	(—)	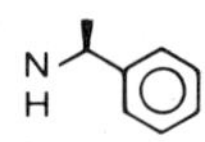		7
56	NO_2	(—)	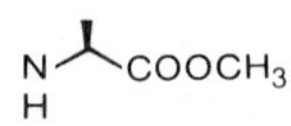		3
57	NO_2	(—)	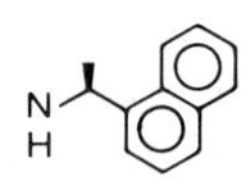		16
5	H	(+ or —)	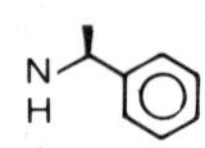	(MBA-NP)	25/25
58	H		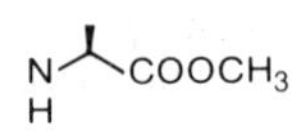		1
59	H		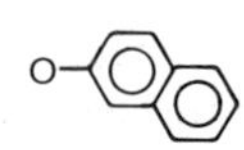		5
60	H	(—)	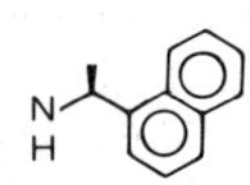		2
61	H	(+,—)	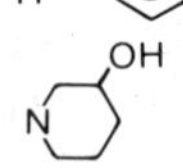		5
62	H	(—)	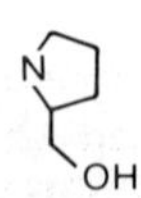	(PNP)	160
63	Me	(—)	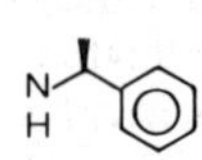		7
64	Me	(—)	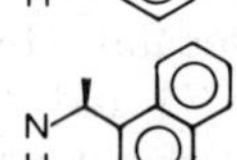		25

have prepared at least twenty more nitropyridines using achiral amines and none of them have any SHG activity whatsoever. The compound MBA-NP 5 prepared from optically active α-methylbenzylamine is some twenty-five times as efficient as urea and has a cutoff at 430 nm. Hydrogen bonding between the amine hydrogen and the nitro group of an adjacent molecule appears to play an important role in the crystal packing of this molecule. Powder measurements also indicate that MBA-NP is noncritically phase-matchable near 0.90μ, a property which makes it very attractive for doubling semiconductor lasers which operate in this wavelength region (46). The very strong SHG activity of PNP 62 containing the optically active prolinol donor is likely due to a variety of factors and deserves some comment. The compound NPP, which contains a benzene ring instead of the pyridine ring of PNP, is even more nonlinear and this activity has been attributed to the near optimal alignment of dipoles in the crystal to fit phase matching requirements (47) and so it is not unreasonable to expect that the alignment in PNP is also highly favorable in this sense. Hydrogen bonding has been found to be an important factor in the crystallization of PNP. X-Ray analysis shows an infinite chain of molecules in the crystal linked end to end by hydroxyl to nitro group interactions. The presence of pyrrolidine as the donor in PNP is also very favorable since its donor ability is second only to the julolidine or tetrahydroquinoline types and certainly much better than dimethylamino (compare R°(pyrrolidine)=-0.63 vs. R°(dimethylamino)=-0.53). A third important structural feature in PNP is the absence of the Birtles-Hampton effect (the reduction of mesomerism by steric interactions) (48) since there are no ortho substituents present for donor or acceptor, and so the full donor strength of the pyrrolidine ring can be involved. In contrast to this an examination of the crystal structure of DAN 42 shows that the donor dimethylamino group is severely distorted due to steric interaction with the adjacent ortho-acetamido group.

A wide variety of other heterocyclic ring systems can conceivably serve as the conjugated backbone in nonlinear organic molecules. We will give examples from preliminary work on two of these, the thiazole and pyrimidine heterocycle derivatives 65-72 in Table VIII. These two heterocycles were chosen because the appropriate haloderivatives are commercially available as starting materials for nucleophilic aromatic substitution. The pyrimidine derivatives are of particular interest since their absorption edges (~400 nm) are shifted hypsochromically an additional 30 nm relative even to the pyridines.

The UV-VIS cutoffs of some model compounds clearly demonstrate a variety of structural effects discussed earlier. Although these spectra were obtained in solution (4×10^{-4}M in EtOH) the trends in transparency of these simple molecules are generally applicable to the crystalline state as well. Two of the more clear-cut comparisons involve the D-A substituted parent pyridine 73-74 and benzene 75-76 molecules (Figure 1). First, the cutoffs of the pyrrolidine substituted aromatics 74 and 76 have a bathochromic shift relative to their dimethylamino substituted analogs 73 and 74 due to the better donor

Table VIII. Thiazole and Pyrimidine Derivatives

R	#	Eff 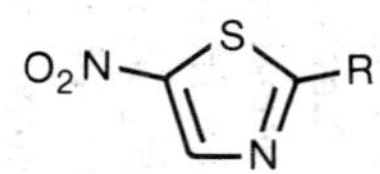	#	Eff 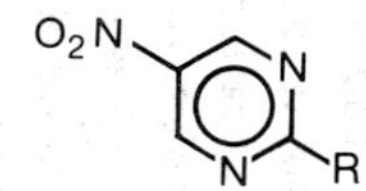
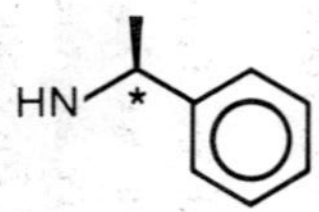	65	8	66	4
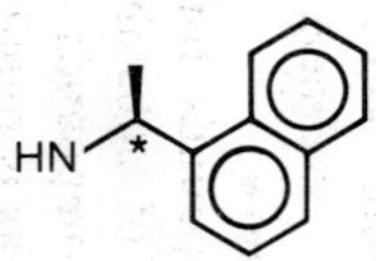	67	0.1	68	2
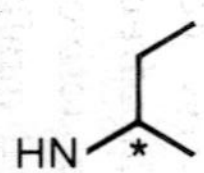	69	weak	70	1
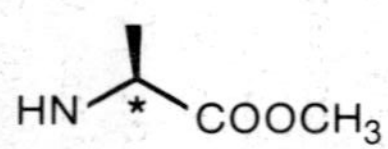	71	3	72	<1

properties of the pyrrolidine. Second, the pyridine derivatives 73 and 74 have a consistent hypsochromic shift in cutoff of some 30 nm relative to their benzene analogs 75 and 76. This is due, in part, to the dipole moment contribution of the pyridine ring which opposes the dipole moment arising from mesomeric interaction of the donor and acceptor. The cutoff behavior of the ortho-methyl substituted systems 77-80 (Figure 2) demonstrates how these relationships can be perturbed by steric interactions. The methyl group itself is a weak donor (R°=-0.10) and is expected to induce a small bathochromic shift of some 5-10 nm. The largest perturbation in cutoff wavelength and slope is found for nitrotoluene 79, in which the sterically demanding dimethylamino group is present. Nitrotoluene 80 which has the less sterically demanding pyrrolidine ring, has almost normal cutoff behavior. Remarkably, the β-picolines 77 and 78 are not nearly as sensitive to steric effects as are the toluenes 79 and 80; this may reflect the difference in steric demands of the nitrogen lone-pair and a hydrogen atom. The overall important conclusion here is that substituents introduced in an attempt to bring about favorable noncentrosymmetric crystallization, even a methyl group, may simultaneously introduce steric influences on cutoff, mesomerism, and thereby effect SHG efficiency as well.

Our efforts to date have concentrated on organic compounds with absorption edges between 0.40μ and 0.50μ. Two of our new materials, MBA-NP 5 (cutoff 0.43μ, 25×urea) and PNP 62 (cutoff 0.475μ, 160×urea) have the largest SHG powder efficiencies known as a function of their respective cutoffs. These are both good materials but there remains significant room for improvement. The most important general finding from our study involves the role that hydrogen bonding can play in obtaining highly efficient SHG in the crystalline state. X-Ray crystallographic analysis of some of our most efficient compounds (e.g., 5, 33 42 and 62) clearly indicates the presence of intermolecular hydrogen bonding which contributes to the ultimate crystal packing of the molecules. We intend to exploit hydrogen bonding in combination with optical activity, if necessary, in our future work. We believe that organic compounds with cutoffs between those of urea 1 and POM 4 (0.20-0.40μ) have been largely neglected and that significant improvements are likely. In addition, a more thorough study of organic dyes (cutoff ≥0.50μ) will undoubtedly reveal nonlinear materials with substantially better efficiencies than are now known.

The SHG active compounds found in Tables II-VIII are just a small fraction of the hundreds of substances we have screened by the powder technique. Of the active compounds only another small fraction will provide crystals of size and quality as is required for optical application. A program in crystal growth as thorough as the effort which has gone into synthesis must be instituted in order to take full advantage of these materials. Overall, it will suffice to say that the promising nonlinear properties observed from such a small fraction of all the possible organic compounds bodes very well for future endeavor.

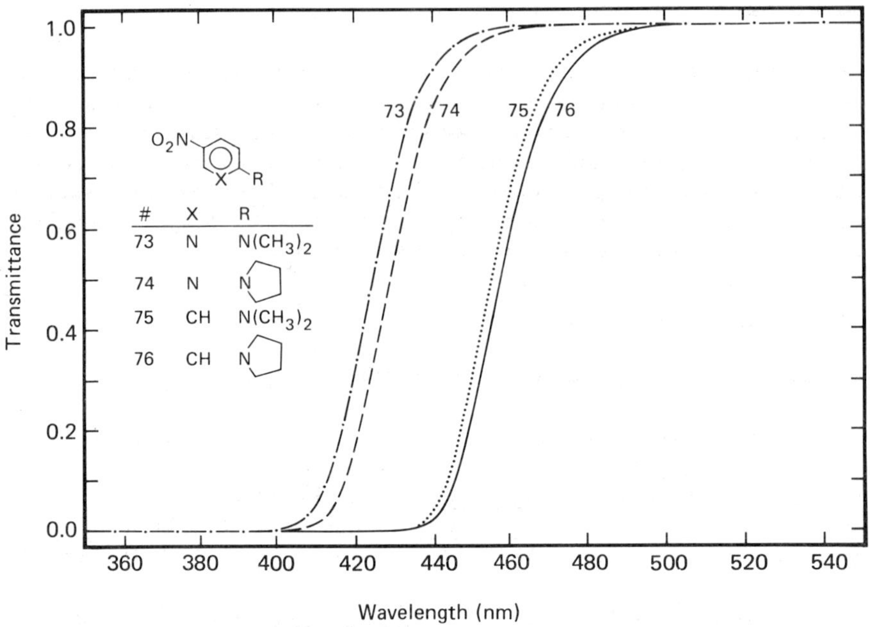

Figure 1. Cutoffs of Model D-A substituted aromatics.

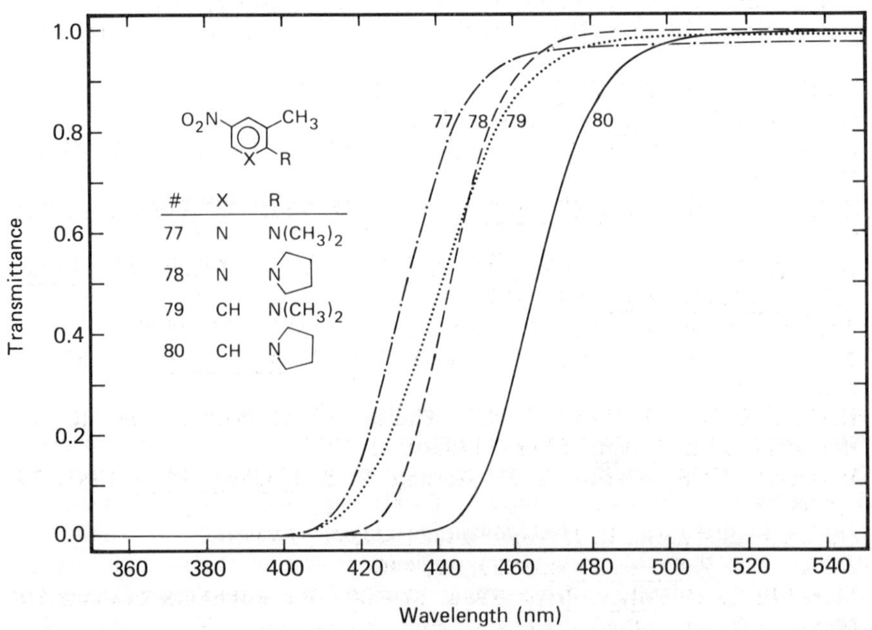

Figure 2. Effects of methyl substitution on cutoffs.

Acknowledgments

The authors wish to acknowledge the collaboration of a number of colleagues who made important contributions to this work: J. I. Crowley and Y. Y. Cheng of IBM Research; visiting scientists G. H. Hewig (Institut fur Physikalische Elektronik, Universitat Stuttgart) and A. Azema (Laboratoire d'Electrooptique, University of Nice); and students P. Bundman (University of Arizona), C. Lehfeldt (Georgetown University), G. Magel (Stanford University) and M. Verdier (San Jose State University).

Literature Cited

1. Franken, P.; Hill, A.; Peters, C.; Weinreich, G. Phys. Rev. Lett. 1961, 7, 118.
2. Armstrong, J. A.;, Bloembergen, N.; Ducuing, J.; Persham, P. S. Phys. Rev. 1962, 127, 1918.
3. Kurtz, S. K.; Perry, T. T. J. Appl. Phys. 1968, 39, 3798.
4. Giordmaine, J. A. Phys. Rev. Lett. 1962, 8, 19; Maker, P. D.; Terhune, R. W.; Nisenoff, M.; Savage, C. M. Phys. Rev. Lett. 1962 8, 21.
5. Halbout, J. M.; Blit, S.; Tang, C. L. IEEE J. Quant. Elect. 1981, QE-17, 513.
6. Chemla, D.; Oudar, J. L.; Zyss, J. Echo Rech. 1981, 103, 3 (English Ed. p. 47).
7. Zyss, J. in Proceedings of the Conference on Optical Properties of Glass and Optical Materials, Part 2 Righini, G. C., Ed.; Florence, Italy, April 6-8, 1982.
8. Curtin, D. Y.; Paul, I. C. Chem. Rev. 1981, 81, 525.
9. O'Loane, J. K. Chem. Rev. 1980, 80, 41.
10. Zernike, F.; Midwinter, J. "Applied Nonlinear Optics"; John-Wiley; New York, 1973.
11. Zyss, J.; Berthier, G. J. Chem. Phys. 1982, 77, 3635 and references 22-28 therein.
12. Jain, K; Crowley, J. I; Hewig, G. H.; Cheng, Y. Y.; Twieg, R. J. Optics and Laser Technology 1981, 297.
13. Zyss, J.; Chemla, D.; Nicoud, J. J. Chem. Phys. 1981, 74, 4800.
14. Twieg, R.; Azema, A.; Jain, K.; Cheng, Y. Y. Chem. Phys. Lett. 1982, 92, 208.
15. Levine, B. F.; Bethea, C. G.; Thurmond, C. D.; Lynch, R. T.; Bernstein, J. L. J. Appl. Phys. 1979, 50, 2523.
16. Lipscomb, G. F.; Garito, A. F.; Narang, R. S. J. Chem. Phys. 1981, 75, 1509.
17. Oudar, J. L.; Hierle, R. J. Appl. Phys. 1977, 48, 2699.
18. Oudar, J. L. J. Chem. Phys. 1977, 67, 446.
19. Meredith, G. R. Polym. Prepr. 1982, 23, 158. We would like to thank Dr. Meredith for providing unpublished data on compound 10.

20. Dulcic, A. Chem. Phys. 1979, 37, 57.
21. "Scientists Seek Nonlinear Organic Materials," Chemical and Engineering News, October 4, 1982, p.18.
22. Oudar, J. L.; Zyss, J. Phys. Rev. A 1982, 26, 2016, 2028.
23. Exner, O. "Dipole Moments in Organic Chemistry", Georg Thieme, Stuttgart, 1975.
24. Edward, J. T.; Farrell, P. G.; Job, J. L. J. Phys. Chem. 1973, 77, 2191.
25. Brooker, L. G. S.; et. al. J. Amer. Chem. Soc. 1951, 73, 5532.
26. Davies, P. L. Trans. Faraday Soc. 1952, 48, 789.
27. Rustagi, K. C.; Ducuing, J. Opt. Commun. 1974, 10, 258.
28. Oudar, J. L.; Chemla, D. S. Opt. Commun. 1975, 13, 164.
29. Oudar, J. L.; Le Person, H. Opt. Commun. 1975, 15, 258.
30. Levine, B. F.; Bethea, C. G. J. Chem. Phys. 1978, 69, 5240.
31. Levine, B. F. J. Chem. Phys. 1975, 63, 115.
32. Dulcic, A.; Sauteret, C. J. Chem. Phys. 1978, 69, 3453.
33. Chapman, N. B.; Shorter, J. Eds., "Advances in Linear Free Energy Relationships," Plenum Press, New York, 1972.
34. Taft, R. W.; Price, E.; Fox, I. R.; Lewis, I.; Andersen, K.; Davis, G. J. Amer. Chem. Soc. 1963, 85, 3146.
35. Katritsky, A. R.; Topson, R. D. Agnew. Chem. Int. Ed. 1970, 9, 87.
36. Fabian, J.; Hartmann, H. "Light Absorption of Organic Colorants," Springer-Verlag, New York, 1980.
37. Kitaigorodsky, A. I. "Molecular Crystals and Molecules," Academic Press, New York, 1973.
38. Kitaigorodsky, A. I.; Mirskaya, K. V. Kristallografiya 1964, 9 634; Kristallografiya 1965, 10, 162.
39. Jacques, J.; Collet, A.; Wilen, S. H. "Enantiomers, Racemates and Resolutions," Wiley-Interscience, New York, 1981.
40. Davydov, B. L.; Derkacheva, L. D.; Dunina, V. V.; Zhabotinskii, M. E.; Zolin, V. F.; Koreneva, L. G.; Samokhina, M. A. JETP Lett. 1970, 12, 16.
41. Jain, K.; Hewig, G.; Cheng, Y. Y.; Crowley, J. I. IEEE J. Quant. Elec. 1981, QE-17, 1593.
42. Bass, M.; Bua, D.; Mozzi, R.; Monchamp, R. R. Appl. Phys. Lett. 1969, 15, 393.
43. Rabinovich, D; Shakked, Z. Acta. Crystallogr. 1974, B30, 2829.
44. J. L. Oudar, Doctor of Science Thesis, University of Pierre and Marie Curie, Paris, 1977.
45. Twieg, R.; Azema, A.; Jain, K.; Lehfeldt, C., manuscript in preparation.
46. Azema, A.; Jain, K.; Twieg, R. "Phase Matching Properties of Nitropyridines," Optical Society of America Annual Meeting, October 18-22, 1982; Tuscon, AZ.
47. Zyss, J., personal communication.
48. Birtles, R. H.; Hampson, G. C. J. Chem. Soc. 1937, 10.

RECEIVED May 16, 1983

4

Nonlinear Organic Crystals: Theoretical Concepts, Materials, and Optical Properties

J. BADAN, R. HIERLE, A. PÉRIGAUD, and J. ZYSS

Centre National d'Etudes des Télécommunications, 196 rue de Paris, 92220 Bagneux, France

Organic molecular crystals exhibit three-wave nonlinear mixing efficiencies one or two orders of magnitude above that of Lithium Niobate. Beyond the basic common molecular features of these crystals (conjugation and intra-molecular charge transfer), different possibilities to achieve a proper lattice orientation (chirality, a vanishing molecular dipole, H-bonding substituent groups) are exemplified by a number of original crystals. Progress in growth (solution cooling and melt) and characterization of organic crystals are reported.

Theoretical Concepts

The fast growing development of optical fiber communication systems has stimulated the search for new highly nonlinear materials capable of fast and efficient treatment of optical signals. Earlier pioneering work had pointed out the higher quadratic susceptibilities of molecular compounds and cubic susceptibilities of polymers, either resonant or non-resonant, and thus extended the domain of research for nonlinear materials from inorganic semiconductors or insulators to organics. In the same period, other potential applications of organics have shown-up such as semiconductivity, superconductivity, photoconductivity, etc... also contributing to the growing interest in these materials.

A common striking feature is the simultaneous appearance in all these branches of the same concept of molecular engineering which can be defined as a voluntary and hopefully predictive chemical action, at microscopic level, on the physico-chemical property of interest and has proved to be in many cases a successfull and unifying approach. The same attempt is also existent in the study of inorganic materials but the innumerable possibilities of organic synthesis, and the higher chemical reactivity of organic

0097-6156/83/0233-0081$08.00/0

materials give to this concept a more general and practical validity. We shall in the following restrict ourselves to the field of 3-photon nonlinear effects.

These effects (1,2) are all driven by the same third-rank frequency dependent nonlinear susceptibility $\chi^2(-\omega_3\ ;\ \omega_1\ ,\ \omega_2)$.d is sometimes preferred for second-harmonic generation (SHG).

Enhancement of χ^2 will lead to improvement (in terms of efficiency per interaction volume) in the following applications : up-conversion in the visible or near U.V. of powerful I.R. laser radiation, frequency modulation of a laser carrier beam, optical parametric oscillation and amplification for solid state infrared tunable coherent devices.

It should be stressed that nonlinear optical effects (particularly optical parametric oscillation) are very demanding in terms of material quality.

In a crystalline medium, the parametric gain (2) Γ^2 is proportionnal to $d^2\ I_p\ n^{-3}$ and the oscillation condition $\Gamma^2\ell^2 > \alpha\ell$ where α is the signal residual absorption (dramatically increased by any crystalline defect), d the efficient phase-matched nonlinear susceptibility, n an average refractive index, I_p the pump intensity (limited by the optical damage threshold) and ℓ the effective interaction length (also limited by any source of crystalline disorientation).

To increase the efficiency and range of applications of parametric oscillation, an effect not yet achieved in an organic material, d should be increased, n lowered (refractive indices of organic compounds are between 1.5 and 2 in the visible and near I.R., to be compared to much higher values of many inorganic nonlinear materials) and the optical damage threshold increased (organic materials will be seen to sustain at least as much radiation power as usual nonlinear inorganic materials). Besides, organic materials of interest being transparent between .5 and 2μm qualify for use as nonlinear devices in optical fiber telecommunications systems.

In molecular crystals, molecules are located at geometrically equivalent sites and related to each other by rather loose intermolecular forces. Intramolecular covalent bonds have binding energies one and two orders of magnitude above that of intermolecular hydrogen and Van-der-Waals bonds. This scaling explains why molecules in crystals retain their individuality and make valid the following additive scheme : calling P the macroscopic polarization, χ^n the nth order crystalline nonlinear tensor, p the molecular dipole α, β, γ, the successive molecular (hyper) polarizabilities, E the optical field and E_ℓ the local field.

$$P = P_0 + \chi^1\ E + \chi^2\ EE + \chi^3\ EEE + \ldots \qquad (1)$$

$$p = p_0 + \alpha\ E_\ell + \beta\ E_\ell\ E_\ell + \gamma\ E_\ell\ E_\ell\ E_\ell + \ldots \qquad (2)$$

Nonlinear processes being basically of intramolecular nature, corresponding terms in the macroscopic and microscopic dipoles expansions can be related by the following tensorial summation (given here for SHG coefficients), following an oriented gas description :

$$d_{IJK} = N\, f_I^{2\omega}\; f_J^{\omega}\, f_K^{\omega}\; b_{IJK} \qquad (3)$$

$$b_{IJK} = \frac{1}{n(g)} \sum_{s=1}^{n(g)} \cos\,[I,i(s)]\, \cos\,[J,j(s)] \cos[K,k(s)] \beta_{ijk} \qquad (4)$$

N is the number of molecules per unit volume (packing density factor), f^{ν} is a Lorentz local field correction at frequency $\nu (f^{\nu} = [(n^{\nu})^2 + 2]/3,\ \nu = \omega$ or $2\omega)$. Although generally admitted, this type of local field correction is an approximation which certainly deserves further investigation. IJK (resp ijk) are axis denominations of the crystalline (resp. molecular) reference frames, n(g) is the number of equivalent positions in the unit cell for the crystal point symmetry group g. b_{IJK} , crystalline nonlinearity per molecule, has been recently introduced (3,4) to get general expressions, independant of the actual number of molecules within the unit cell (possibly a (sub) multiple of n(g)).

Expression (4) evidences the combined influences of molecular and crystalline structure on the enhancement of the macroscopic optical nonlinearity. The molecular design must contain specific features so as to optimize β_{ijk} while the crystalline structure (i.e. the arrangement of molecules in the unit cell), appearing in the form of projection factors in expr.4, must eventually promote at the macroscopic level, optimized responses of individual molecules. Advances have been reported in the understanding of relations between molecular hyperpolarizabilities and molecular structure, leading to a so-called "molecular engineering" approach and the design of "tailor made" efficient nonlinear molecules. However, no general theory is presently available on the prediction of the structure of the crystal, from the sole knowledge of the molecular structure. However, a general and quantitative definition of the optimal molecular orientations, valid for all crystal point groups and efficient molecular units has been recently proposed (3, 4). Designing a molecular compound with a higher quadratic macroscopic susceptibility requires first the definition of an "optimized" molecule in the first step and capability in a second step of organization in a crystalline lattice of suitable symmetry as detailed below.

Molecular engineering

Molecular features responsible for the enhancement of three photon effects were originally identified in a rather empirical way, by scanning hundreds of organic compounds (5, 6) using the now standard second-harmonic generation (SHG) powder test (7). The first basic feature is the presence of highly polarizable electronic configurations in the molecule. Such is the case in highly conjugated systems either linear (bond alternation such as in polyethylenic chains) or cyclic (aromatic systems). It is not sufficient to ensure a nonlinear response : due to the odd rank of tensor, the molecular system must not be centrosymmetric. A third feature is the occurence of intramolecular charge transfer such as happens in paranitroaniline.

Some substituents induce remarkably different electronic behaviors on the same aromatic system (8). Let us consider, for example, the actions of substituents on an aromatic electron system. Some substituents have a tendency to enrich their electronic population (acceptors), while others will give away some of it (donors). Traditionaly, quantum chemists used to distinguish between long range (mesomeric) effects, mainly π in nature, and short range (inductive) effects, mainly σ. The nonlinear behavior of a monosubstituted molecule can be accounted for in terms of the π electron dipole moment. Examples of donor and acceptor substituents can be seen on figure 1.

When located at opposite ends (or at conjugated positions) in a molecular system, a donor and an acceptor do more than simply add up their separate effects. A cooperative phenomenon shows up, involving the entire disubstituted molecule, known as charge transfer (C.T.). Such compounds are colored (from pale yellow to red, absorption from 3,000 to 5,000 Å) and show high U.V. absorption oscillator strength. "Figure 2" helps understand the enhancement of optical nonlinearity in such a system.

Generally, the transparency of a molecular π electron system narrows with increasing conjugation (bathochromic effect) while its nonlinear efficiency increases. The urea molecule is a small conjugated molecule transparent up to 2000 Å with a low β value : $1.3\ 10^{-30}$e.s.u. (10). For 4-nitro 4'-dimethylaminostilbene the situation is opposite : absorption occurs at 5,000 Å and β is 450.10^{-30}e.s.u. (11).

In order to measure molecular hyperpolarizabilities the now standard D-C induced SHG experiment is used (12). Although it would be more suitable to work in the gas phase to minimize molecular interactions, high molecular weights (low vapour pressure) and chemical decomposition processes make it hardly feasible for the molecules of interest.

In solution, although solute contributions can generally be singled out, difficulties arise sometimes : solvent-solute interactions may induce a shift of the solute absorption and consequently of its susceptibility or hydrogen bonded molecular complexes may modify the liquid structure. This situation has been studied both theoretically and experimentally by Zyss and Berthier (10) and by Ledoux and Zyss (13) in the case of urea derivatives in various solvents and in crystal showing the importance of environment considerations and thus the limitations of an oriented gas model for crystals.

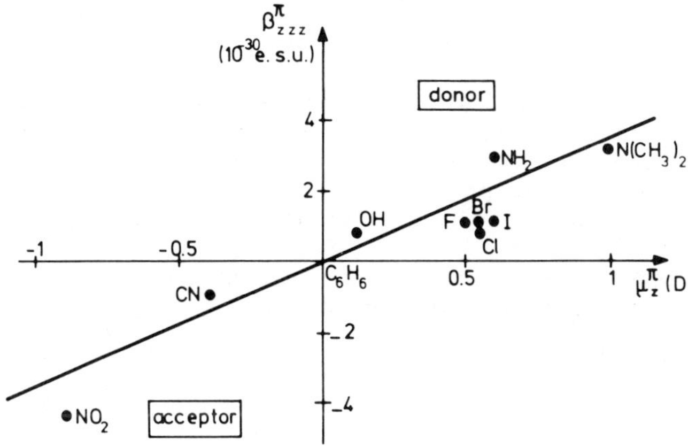

Figure 1. Nonlinearity of some monosubstituted aromatic molecules in terms of their dipole moments (π electron contributions).

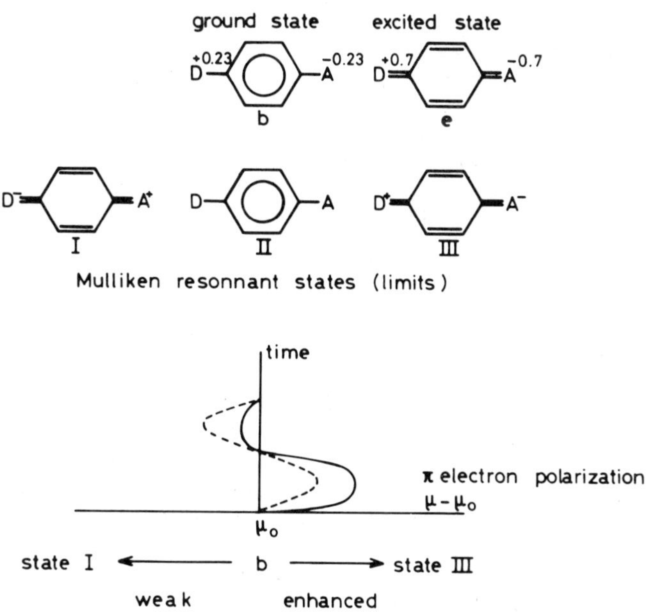

Figure 2. Origin of the nonlinearity of charge-transfer molecules for a two-level-like disubstituted aromatic molecule. The donor (D) and acceptor (A) cooperate to distort the linear response (dotted lines) of a nonsubstituted benzene ring (below).

We shall review some molecular models, at different levels of approximation, which help theoretically support the previous results.

A simple model, known as the internal equivalent field model (9) (see "Figure 1") accounts for the different β values of monosubstituted benzene derivatives R-X, and relates them to the respective donor or acceptor strength of X. It is assumed in this model that the substituent action on the π electrons of the ring is equivalent to that of a D-C field E_0 with direction and intensity related to the substituent electronegativity. Identifying dipole expansions of R-X in presence of E and of R in presence of $E_0 + E$ yields :

$$\mu^{\pi}_{R-X} = \alpha^{\pi}_{R} E_0 \qquad \beta_{R-X} = 3(\gamma^{\pi}_{R}/\alpha^{\pi}_{R})\mu^{\pi}_{R-X} \tag{5}$$

Although satisfactory for weak donor-acceptor interactions, this model does not hold in the case of strong charge transfer such as happens in some disubstituted benzene derivatives. A simple additive assumption falls short of the experimental values for the three nitroanilines in the para, ortho and meta positions (14). Discrepancies between an additivity assumption and experimental values are seen to increase with the degree of conjugation induced by the substituent relative locations.

A quantum mechanical expression for β can be straight forwardly obtained from perturbation theory (with dipolar perturbation coupling of the molecule with the electromagnetic field).

In the case of a charge transfer molecule such as paranitroaniline, only two electronic states contribute significantly in the visible range. The rest of the spectrum contributes a small, if not negligible, additive background. In the latter case, β reduces to :

$$\beta_{ijk} = \frac{1}{2h^2}\left\{\frac{\delta_i m_j m_k}{\omega_1^2 - \omega^2} + m_i(m_j\delta_k + \delta_j m_k)\frac{\omega_1^2 + 2\omega^2}{(\omega_1^2 - 4\omega^2)(\omega_1^2-\omega^2)}\right\}$$

where $\underline{\delta}$ is the difference between the dipole moments of the excited and ground states, $\underline{m}$ is the transition dipole connecting these states, $h\omega_1$ the gap $h\omega$ the fundamental photon energy.

If $\underline{a}$, $\underline{b}$ are two unit vectors arbitrarily chosen in the $(\underline{\delta},\underline{m})$ plane, $\underline{\beta}$ is seen from the previous expression to be a two-dimensional tensor with only four components β_{aaa}, β_{bbb}, β_{abb}, β_{baa}. This situation will be more thoroughly studied in the following for methyl-(2,4-dinitrophenyl) α-aminopropanoate (MAP) crystals.

Quantum mechanical models at different levels of approximation have been successfully applied to compute molecular hyperpolarizabilities. Some authors have attempted a complete determination of the U.V. molecular spectrum to fill in the expression of β (15, 16). Another approach is the finite-field perturbative technique (17) demanding the sole computation of the ground state level of a perturbated molecule, the hyperpolarizabilities being derivatives at a suitable order of the perturbed ground state molecule by application of the Hellman-Feynman theorem.

Crystal Engineering and Efficiencies of 3-methyl 4-nitropyridine I-oxide (POM), MAP and N-4-nitrophenylprolinol (NPP)

The second step of the nonlinear material optimization process will be detailed in the cases of three efficient nonlinear materials recently developed at CNET : POM, MAP and NPP. For all these materials specific features were implemented at the molecular level so as to act on the crystalline structure in the required direction (i.e. away from centrosymmetry and toward an "optimal" crystalline structure which has been theoretically defined (3, 4) and actually almost reached in the case of NPP.

These features, made compatible with the previously detailed electronic requirements, are respectively : a vanishing ground state dipole moment requirements in the case of POM (18), chirality in the case of MAP (19), chirality and hydrogen-bonding in the case of NPP (20). Characterization and growth of these materials will be detailed further on.

The interest of defining an efficient nonlinear molecule with the additional feature of a vanishing dipole moment lies in the following expected potential advantages : first, in the absence of stronger H-bond intermolecular interactions, intermolecular dipole-dipole electrostatic interactions tend to locate molecular dipoles in opposite directions, thus favoring a centrosymmetric lattice. Cancelling-out dipole-dipole interactions does not guarantee the absence of centrosymmetry (as chirality does) but should result in a structure, made more "sensitive" to other influences (such as minor molecular substitutions), potentially able to prevent centrosymmetry and otherwise masked by stronger electrostatic forces. Secondly, disruptive molecular associations in solution (solvent-solute or solute-solute) are less likely to appear resulting in easier desorbtion of the solute during the crystal growth process in solution and more favorable initial conditions toward the growth of large, high quality samples.

However, definition of an efficient molecule for N.L.O. with a vanishing dipole moment requires solving a serious but not intrinsic contradiction : most conjugated charge-transfer molecules have significant dipole moments owing to the non-negligible resonant contribution of a highly polar $A^{-} - 0 - D^{+}$ structure to the molecular ground state (see for example Mulliken expresions of the dipole given in 21). In POM the localized dipole moment of the N-oxide semi-polar bond (4.2 D) opposes and cancels-out that of nitrobenzene resulting in a negligible value of the total dipole. Besides, in an heterocyclic molecule the N-oxide group may act as well as a donor or an acceptor group depending on the electronic nature of the para substituent : in the present case, the acceptor property of the nitro group promotes the donor nature of the N-oxide bond.

With the addition of a methyl group in the 3 position which does not interfere with the basic charge transfer process, the resulting crystalline structure is $P_{2\ 2\ 2}$.

Crystalline samples of large size (few cm^3) have been grown in solution and characterized by X-ray Lambot method showing good crystalline quality. SHG coefficients have been measured (18) : $d_{14} = d_{25} = d_{36} = 23 \pm 3.10^{-9}$ e.s.u. (i.e. nearly 14 times d_{36} of KDP). 50 % efficient SHG conversion efficiency was achieved on 5mm yielding 40 MW/cm^2 of a .53 μm doubled Q-switch YAG output. Damage threshold is 50 MW/cm^2 at .53 μm. Pumping a singly resonant oscillator at degeneracy (ω_p = .53 μm, $\omega_i = \omega_s$ = 1.06 μm) with 40 MW/cm^2 pump intensity results in the favorable ratio $\Gamma^2 \ell^2_{eff}/2\alpha_\omega \ell$ =2,67 (where Γ is the parametric gain, ℓ_{eff}= 3.5 mm, ℓ = 5 mm, α_ω =0.77cm^{-1} at 1.06 μm). A much higher gain should be obtained by raising the pump wavelength and taking advantage of a higher damage threshold to pump more efficiently. A specific feature of curve n°1 of figure 3 is to be noted : above 1.4 μm SHG phase-matching geometric conditions are seen to depend weakly on the fundamental wavelength. This should make possible the nonlinear treatment of broad optical pulses (such as frequency doubling of picosecond pulses) without spectral narrowing, or the doubling of a tunable I.R. source with non critical adjustement of the geometry. Electrooptic coefficients have been measured by Sigelle and al.(22) (r_{41} = 3.6,r_{52} = 5.1,r_{63} = 2.6 in 10^{-12} m/V) and related to the SGH coefficients by a two-level quantum model of the molecule and adequate dispersion of local field factors. As for MNA (23), the major part (i.e. 75%) of the electrooptic coefficients originates from purely intramolecular electronic contributions.

In the case of MAP, the concept of chirality was used so as to prevent centrosymmetry : a chiral molecule cannot be superimposed on its image by a mirror or center of symmetry so that a crystal made only of left or right-handed molecules can accomodate neither of these symmetry elements. This use of the chirality concept ensures exclusion of a centrosymmetric structure. However as we shall see in the following, the departure of the actual structure from centrosymmetry may be only weak, resulting in limited nonlinear efficiencies. A prerequisite to the introduction of a chiral substituent in a molecule is that its location should avoid interfering with the charge-transfer process.

For MAP, the molecular nonlinear behavior has been shown to be very similar to that of 2,4-dinitroaniline (resp. 22 and 21.10^{-30} esu for β_μ (projection of the vector part of β on dipole $\vec{\mu}$). MAP offers different phase-matching possibilities (see "figure 4") covering continuously its whole transparency domain (from .5 to 2μm). The most efficient one (SHG at 1.06 μm or inverse degenerate parametric emission pumped at .53μm) has a figure of merit 15 times higher than that of $LiNbO_3$. Optical damage threshold is $150MW/cm^2$ at .53μm and $3GW/cm^2$ at 1.06μm (10 ns duration Q-switched YAG laser). 30 % SHG conversion was obtained in a 1 mm thick sample and parametric oscillation (Γ = 14 cm^{-1} at degeneracy) should be feasible on a similar interaction length when crystalline samples of suitable quality are available. More details are to be found in 19.

MAP has a non-centrosymmetric structure (24) P_{2_1} leading to four non-vanishing crystalline nonlinear coefficients. According to the previously defined two-level model, there are also four non-vanishing molecular nonlinearity coefficients which are linearly related to the crystalline coefficients. (3)

However, its was found possible to infer all four microscopic tensor coefficients from macroscopic crystalline values and this impossibility could be related to the molecular unit anisotropy. It can be shown that the molecular unit anisotropy imposes structural relations between coefficients of macroscopic nonlinearities, in addition to the usual relations resulting from crystal symmetry. Such additional relations appear for crystal point group 2,m and 3. For the monoclinic point group 2, this relation has been tested in the case of MAP crystals, and excellent agreement has been found, when taking into account crystal structure data (24), and nonlinear optical measurements on single crystal (19). This approach has been extended to the electrooptic tensor (4) and should lead to similar relations, when the electrooptic effect is primarily of electronic origin.

It is also of great interest to infer the individual molecular hyperpolarizability components from measured nonlinear optical data. We have therefore studied for each crystal point group the possibility of such a determination, taking into account the two-dimensional response of molecular units. We have shown that even in the favorable case of low crystalline symmetry ensuring the presence of at least four independent non-zero nonlinear coefficients, one cannot always infer all four molecular hyperpolarizability components from the sole experimental data on crystals. The reason is that the linear system connecting microscopic and macroscopic nonlinearities is singular, which clearly appears in adequately defined reference frames. This singularity is proved to originate from the crystal symmetry and, in fact, also gives rise to the above mentioned additional structural relations. As an

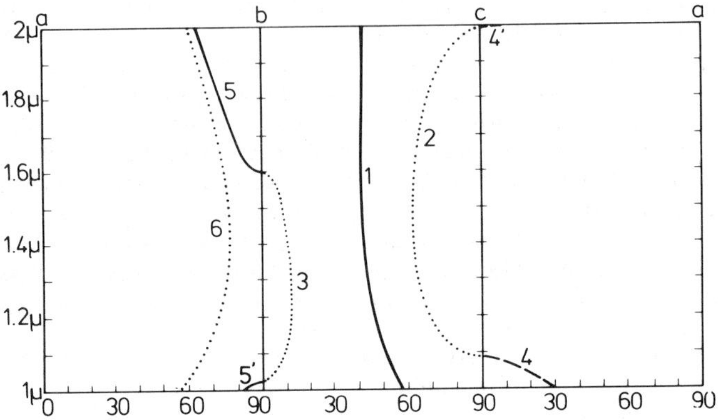

Figure 3. SHG phase-matching configurations in the three principal planes of POM as a function of fundamental wavelength (---) for Type II phase-matching and (—) for Type I phase-matching [(···) theoretical possibilities forbidden by symmetry].

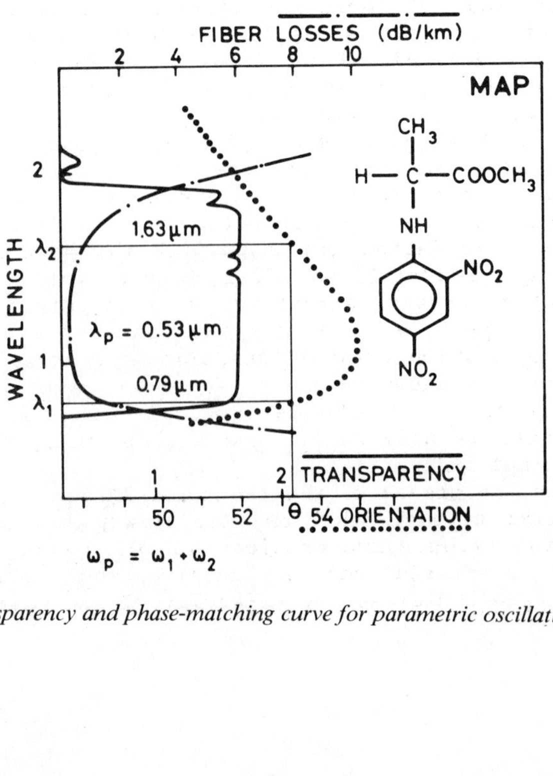

Figure 4. Transparency and phase-matching curve for parametric oscillation of MAP.

point group 2 only three out of four hyperpolarizability components can be determined by SHG data on the crystal. The additional data from electric-field induced SHG measurements in solution has yielded the missing component. As a result, the two-dimensional tensor has been fully determined, and its variation is displayed on figure 5 referenced in an orthogonal system of axes, rotating within the aromatic plane. This result shows that a one-dimensional model would be inadequate for describing β, which is not surprising since three radicals are involved in the charge-transfer interactions. However, the β coefficient is the largest one, which indicates that the charge transfer is predominantly from the amino group to the nitro group in para position, rather than towards the nitro group in ortho. Similarly the linear polarizability has been determined and shown to be isotropic within the conjugation plane. As a consequence, the intersection X of the two molecular planes in the unit cell has to be a dielectric axis which is in agreement with experiment. Semi-empirical INDO calculations (17) have confirmed these features.

This analysis is a convenient basis for discussing the optimization of molecular orientations, beyond the well-known requirement of no center of symmetry. For example, in the case of MAP, we show that the proximity of the predominant molecular charge transfer axis a to the dielectric axis X limits the macroscopic coefficient values. Optimization of the molecular orientation (e.g. with $\Theta = 100°$ or $48°$ instead of $14°$ in actual MAP (see figure 5) would ensure an increase of these coefficients by almost one order of magnitude. Finally, the situation where the molecular nonlinear response is one-dimensional can be studied. In that case the molecular orientation enters through a simple projection factor in the relation between microscopic and macroscopic nonlinearities (e.g. $\sin^2\Theta\cos\Theta$). We have considered, for each point group symmetry, which molecular orientation maximizes this projection factor for a phase-matched interactions. This leads to a classification (shown on "figure 6") of the various point groups with respect to the maximum value this factor can reach with the best possible orientation. These considerations offer a quantitative framework for comparing actual crystal structures with optimized ones.

In that respect we have recently discovered a new compound, namely NPP (20), which crystallizes also in a monoclinic P_{2_1} structure. Moreover, as shown on figure 7 the angle between the charge-transfer axis of any of the two molecules in the unit cell with the binary axis is only at 4° from the optimum orientation (respectively 58 and 54°).

The molecular planes of the two molecules in the unit cell are nearly parallel to each other and the binary axis which makes the actual 2 point group symmetry close to mm2 symmetry. The crystal is thus near to uniaxial and the "quasi-optical axis" close to the direction perpendicular to the molecular planes (assuming that the

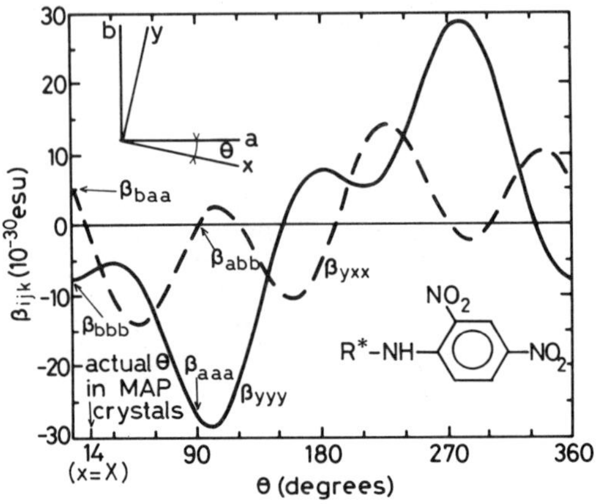

Figure 5. Variation of the molecular β components upon rotation of MAP molecules in the ab plane to give evidence for optimal θ values.

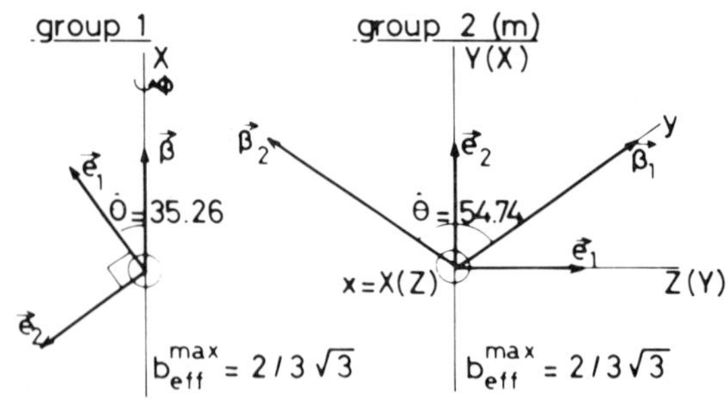

group	b_{eff}^{max}
1, 2, m, mm2	2/3 √3
$\bar{6}$2m, $\bar{6}$, 3, 3m, 32	1/4
222, 6mm, 6, 4mm, 4, $\bar{4}$2m, 23, $\bar{4}$3m	1/3 √3
$\bar{4}$	0

Figure 6. Classification of the noncentrosymmetric crystal point groups by decreasing value of the maximal efficient phase-matchable nonlinear coefficient per molecule (b_{eff}^{max}).

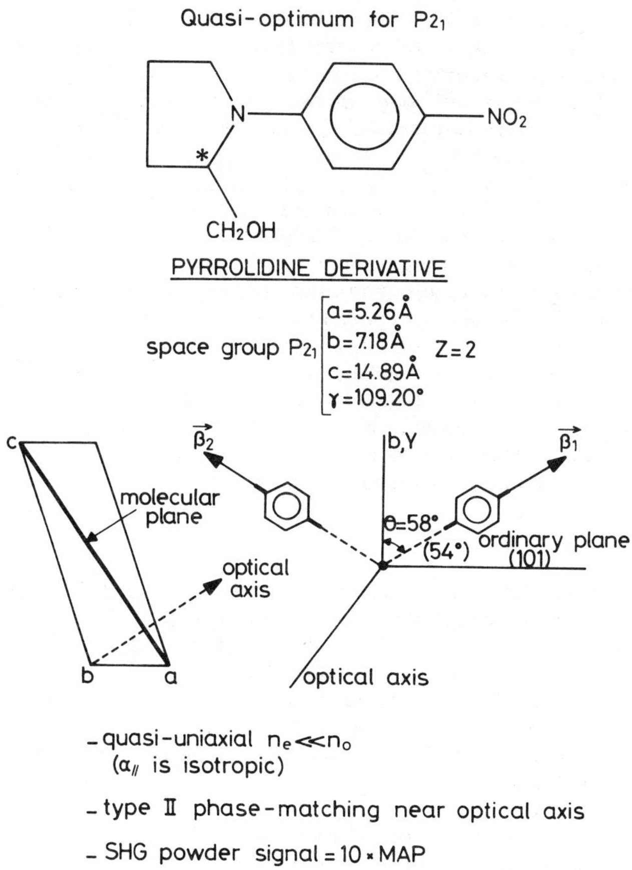

Figure 7. Main crystallographic and nonlinear features of NPP.

molecular polarizability is isotropic in the molecular plane as for MAP).

The analysis of intermolecular distances has shown the prevailing influence, toward the determination of the crystal structure, of intermolecular H-bondings via the prolinol alcohol group. H-bonds, owing to their higher energies will impose the crystalline structure, their influence prevailing over that of dipole-dipole interactions which favor centrosymmetric lattices : a detailed crystallographic (20) study evidences intermolecular H-bonding between molecules located within the same (101) plane and linking the alcohol oxygene of one molecule to one of the nitro group oxygens of the next molecule (average O-O distance 2.86 Å). We have actually observed, on a number of powders, a greater occurence (above 50 %) of non-centrosymmetric structures where alcohol groups were present at the molecular level, than in the absence of H-bonding groups (below 10 %). A possible interpretation for that statistics is that hydrogen bonds which do not favor centro or non-centrosymmetric structures dominate dipole-dipole interactions which definitely favor antiparallel centrosymmetric dimerisations. As could be expected from calculations, knowing the crystalline structure of NPP and assuming that the molecular susceptibility compares with that of nitroaniline, we have found a second harmonic response, on powders, one order of magnitude above that of MAP, thus two orders of magnitude above $LiNbO_3$. The transparency of this new compound compares with that of MAP.

Crystal Growth

Let us consider the crystal growth of the three previous compounds which have retained our attention by their high nonlinear optical susceptibilities (POM, MAP, NPP) plus an additional one : R(+)N-methylidene (gem-carbethoxiphenyl) amide of 4-N-dimethylamino- benzylidene cyanacetic acid -AMA, of potential interest but not yet investigated from the optical point of view. We look successively into their synthesis, purification, crystal growth and characterization.

Synthesis. Because of their molecular originality, three of these compounds have been synthetized. AMA is the condensation result of an acid chloride with chlorhydrate of ethyl R(+)α-phenylaminoacetate according to the reaction process shown "figure 8". Its complexity required a considerable effort to obtain the large quantity of product necessary for crystal growth. The follow-up of the kimetic of the reactions by gas chromatography (GC) and nuclear magnetic resonance (NMR) has permetted the identification of the optimal conditions for the preparation of the crude product.

For MAP, the corresponding acid is prepared by the Sangers method (25) and esterified by methanol or placed in reaction with diazomethane.

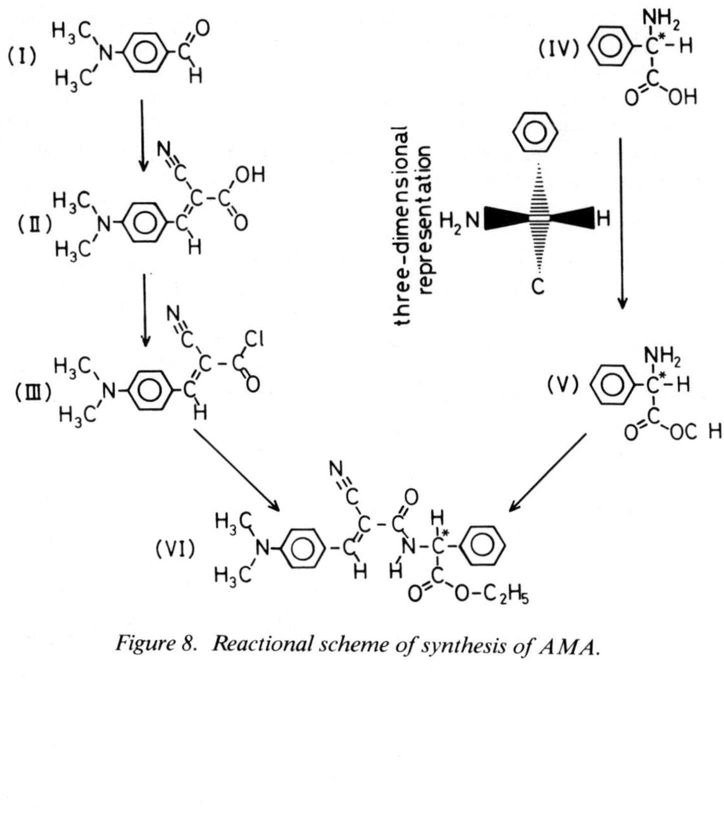

Figure 8. Reactional scheme of synthesis of AMA.

The condensation of 1-fluoro 4-nitrobenzene with prolinol gives NPP. Only POM is a commercial product.

The achievement of the corresponding monocrystals of sufficient optical and crystalline quality is made possible only after very thorough purification. Chemical impurities are known to disturb the crystal lattice through the occurence of twins, veils dislocations, rounding-off of faces ultimately quenching further growth. Any crystalline defect dramatically increases the residual absorption coefficient and lowers the optical damage threshold.

Purification. This problem is related to the degree of purity of the original commercially available compounds and to the elimination of undesirable compounds existing or formed during synthesis. It is advisable to choose the reactions which give a restricted or null number of mean product isomers. The purity of necessary reagents and solvents should be as high as possible. Finally, we must ensure the complete elimination of the excess of reagents and by-products which are always present in greater or lesser quantities.

The crude product is at once purified according to the usual methods of organic chemistry. Then it is essential to lower the impurities rate to a threshold compatible with crystal growth and especially with the use of the materials for nonlinear optics (10 to 100 parts per million).

Repeated recrystallization in an adequate solvent allows a suitable elimination of metallic impurities (see "table I"). Such impurities are known to be very damaging to the optical quality of inorganic crystals as exemplified by Fe impurities in $LiNbO_3$.

Table 1-Residual metallic impurities in MAP and POM at different stages of purification.

		Impurities (ppm)					
		Si	Mg	Fe	Al	Cu	Ca
MAP	crude	55	32	12	25	20	165
	recrystallized	20	6	-	-	8	5
	evaporated	.4	22	-	-	7.5	-
	monocrystal	2	.5	-	-	-	-
POM	crude	15	21	-	10	2	23
	recrystallized	2	6	-	1	-	-
	growth solution	2	2	-	-	-	-
	monocrystal	-	12	-	-	-	5

Any organic impurities, more soluble than the mean product in the respective solvents, remain in solution and are thus removed. At this stage, AMA and POM revealed a sufficiently high degree of purification so that it was not judged necessary to use another method. This is not the case with MAP. Differential thermal analysis (DTA) of the material recrystallized five times in methanol shows the persistent presence of impurities (several percents) which are removed by the following process (see "figure 9").

MAP is one of the organic compounds which have a sufficiently high vapour pressure at $10^{-5}/10^{-6}$ mm Hg to make possible vacuum evaporation in a temperature gradient and permit the collection of a highly purified product in a specific zone. It is advisable not to heat the material too much above its melting point.

Zone refining is used to further purify the material. Enclosed in a sealed tube under inert gas at low pressure, it is subjected to a fusion/recrystallization sequence by displacement of the tube in a series of furnaces interspaced by coolers. The impurities which have a segregation coefficient differing from unity (smaller or greater) accumulate in either the top or bottom of the ingot. The central part can attain a very high degree of purity depending on the number of times it passes through the apparatus. This technique is successfully applied to MAP and NPP.

The different purification stages are controlled by characterization means which are described further.

Physical and chemical properties. Only the physical and chemical properties which have an effect on the crystal growth of these materials are mentioned. The physical properties are grouped in table II.

Table II-Physical properties of the four organic compounds.

	AMA	MAP	POM	NPP
M.P. (°C)	136.35	80.90	136.25	115.93
H_f (Kcal/mole)	20.27	13.80	37.30	
B.P. (°C)	decomp.	decomp.	decomp.	decomp
α_D^{λ}				
at λ = 6328 Å	+13.7498			
5890	+25.6475	-10.70		
5780	+30.3222	-13.34		
5460	+50.2211	-25.30		
crystal class	monoclinic	monoclinic	orthorhomb	monoclinic
Space group	$P2_1$	$P2_1$	$P2_12_12_1$	$P2_1$
Z	2	2	4	2
a	20.963	6.829	21.359	5.257
b (Å)	5.244	11.121	6.111	7.178
c	9.392	8.116	5.134	14.893
β (°)	99.47	95.59		109.20

M.P. melting point ; H_f heat of fusion; BP Boiling Point ; α_D^{λ} specific rotation in degree at λ(Å) ; Z, a, b, c, β lattice parameters.

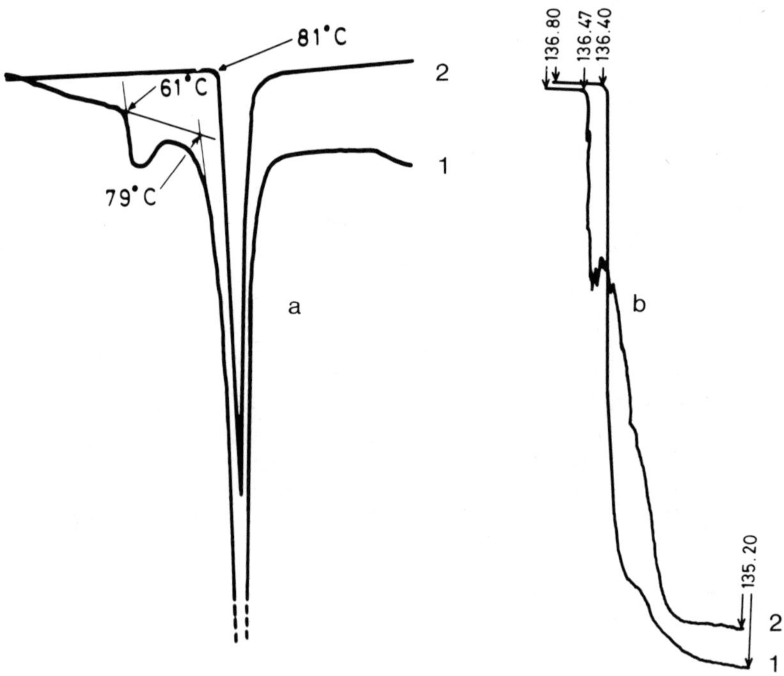

Figure 9. Differential thermal analysis of recrystallized (1) and evaporated (2) MAP (a), and compared melting curves of AMA (1) and POM (2) (b).

These compounds present a greater or lesser degree of instability with regard to air, light, water and heat. The most stable is POM which reacts only in contact with acetylacetone and water. AMA degrades in presence of nitrated or chlorinated solvents. The effect of heat, light, water and methanol on methyl alaninate leads to degradation and to a number of polycondensation compounds so that we may assume that they also appear in the case of MAP under similar conditions (26-32) (see "figure 10"). The solution labilities in most organic solvents have been established for MAP as well as for NPP, preventing use of any crystal growing technique involving such solvents.

Crystal Growth. Depending on the materials, two types of crystal growth method have been used : solution for AMA and POM, melt for MAP and NPP.

Solution growth. The well-known growth technique by temperature decrease of a saturated solution of the compound in a suitably chosen solvent has been employed for AMA and POM. The emphasis has been put on the thermal stability of the baths (better than one hundredth of degree) and on the elimination of stray vibrations.

The solution is stirred using the accelerated seed rotation technique (ASRT) with a maximal speed of 60 rpm and a cycle of 75 seconds. The growth seeds are obtained spontaneously by solvent evaporation at room temperature and in a later stage for POM, by cutting (.82 x .82 x 1 or 2 mm) parallelepipeds along the crystallographic directions in a monocrystal. The solubility curves are reported on "figure 11". The typical conditions of crystal growth are mentioned in table III.

Table III-Growth conditions of AMA and POM by a temperature decrease method in solution

	AMA		POM	
Solvent	Eth/ACN 57 43		Me Acetate	
V	.8		1.2	
T	19.1°C		40.6°C	
T	16.4°C		34.47°C	
Time (days)	34		14	
ΔΘ	.08°C		.05°C	
Size (mm) and weight (g)	10x8x4	.359	6x7x9	.581
	21x6x4	.505	37x10x21	1.986
	15x6x3	;261	7x15x15	2.242

V solution volume (inℓ) T initial T final temperature ΔΘ Medium daily decrease - Eth. ethanol,ACN acetonitrile,Me acétate methyl acetate.

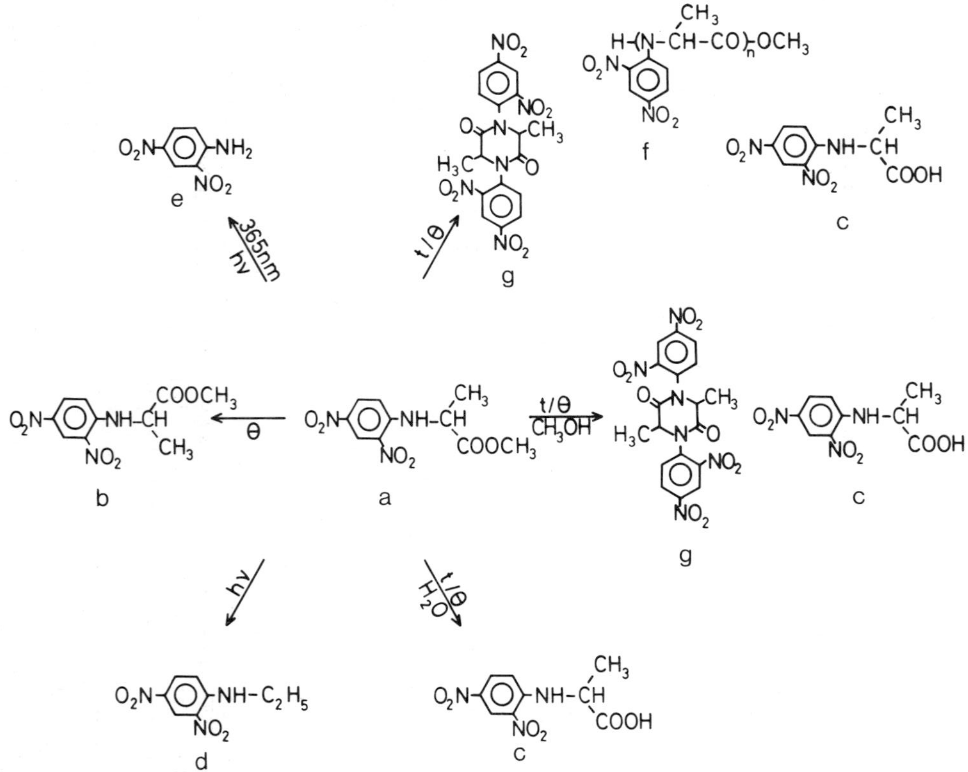

Figure 10. Possibilities of degradation or polycondensation reactions of MAP. Key: a, MAP; b, optical isomer; c, DNP-alanine; d, DNP-ethylamine; e, 2-nitroso-4-nitroaniline; f, DNP-alanyl-(DNP) α amino-propanoate of methyl; and g,1,4-bis(DNP)-3,6-dimethyl-2,5-dioxopiperazine. [*(DNP) = 2,4-dinitro-phenyl.*]

Growth from the melt. The growth method which is used for MAP is derived from the Bridgman-Stockbarger technique. Nevertheless, it is distinguishable from it by the requirement to keep only a minimum quantity of material melted in order to limit thermal degradation or polycondensation over time. This requires a compromise between the temperature and the growth rate of the monocrystal to assure good crystalline and optical quality. The apparatus is made of three adjacent zones, at different temperatures, the melting of MAP taking place in the intermediate zone. The temperature gradient near the melting point should be sharp, typically 10°/cm. The displacement rate of the ampoule containing MAP is 3 to 6 mm/day. The resulting monocrystal has a diameter between 8 and 10 mm and length between 4 and 7 cm. The same growth technique is used for NPP.

Characterization. It is both important for crystal growth and for checking the purification of materials. It takes place at every stage, from synthesis to the monocrystal. There are two aspects : the control of material purity before crystal growth and the control of the crystalline quality of monocrystals.

Control of the chemical purity. Several means are used for this control : Microanalysis, NMR, and UV, IR and X spectroscopies are employed for the identification of synthesized powders. Purification follow-up may be achieved through thermal behavior study with the help of melting point measures, DTA and melting curves (see "figure 9"). Spark emission spectroscopy, GC and high performance liquid chromatography (HPLC) are used for the determination of the quantities of residual impurities. GC shows the high degree of purity reached by AMA and POM after they have been recrystallized many times. HPLC is only one of the means used to detect traces of impurities in MAP, estimated at several tens of ppm in the evaporated material. This method reveals that impurities are attributable more to thermal degradation or polycondensation during growth than to a concentration of impurities distributed in the rod before its crystallization (see "figure 12"). The separation of impurities at each stage by HPLC should soon permit us to identify them by coupling HPLC with mass spectroscopy.

Control of the crystal quality. X-rays diffraction or reflection on the monocrystals reveals the molecule position in the crystal lattice. Laüe diagrams are especially used to index the faces of monocrystals grown in solution (see "figure 13"). For MAP crystal, the absence of faces and the low symmetry of the crystal system has required use of the conoscopic images technique which allows the determination of the orientation of the binary axis with respect to the displacement axis of the crystal during its growth. Conoscopic fringing has permitted determination of the orientation of the XZ plane containing the two optical axis and subsequently,

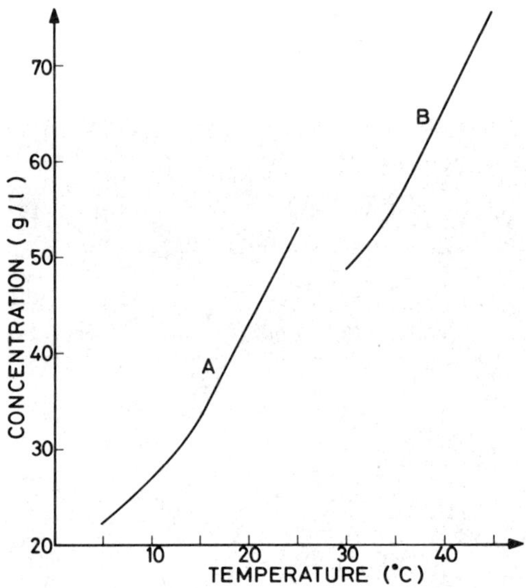

Figure 11. Solubility curves of AMA (A) and POM (B).

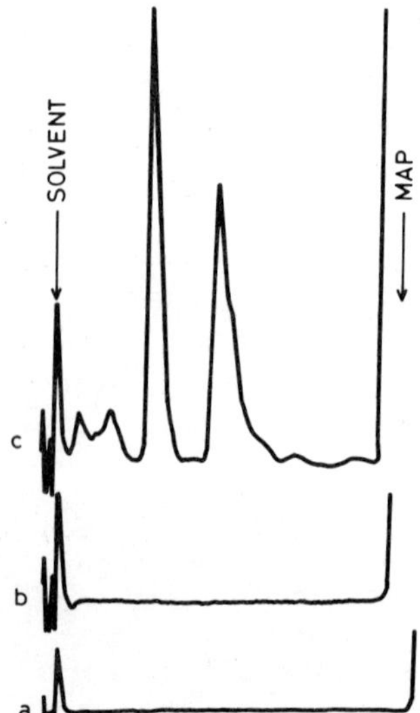

Figure 12. High performance liquid chromatography of various cuttings of MAP ingot. Key: a, top; b, central part; and c, bottom.

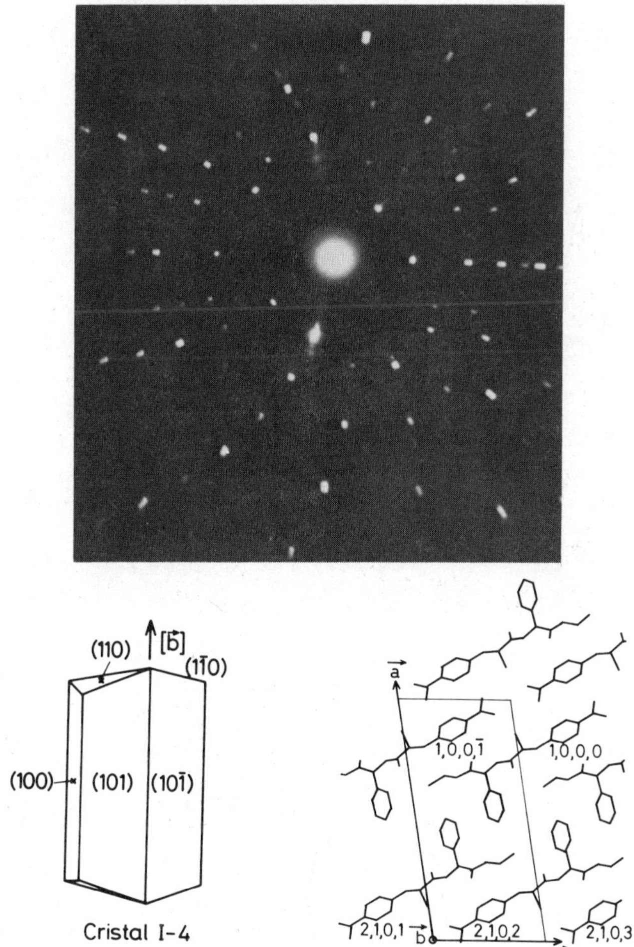

Figure 13a. x-Ray characterization of AMA. Key: top, Laue diagram; bottom left, morphology; and bottom right, structure.

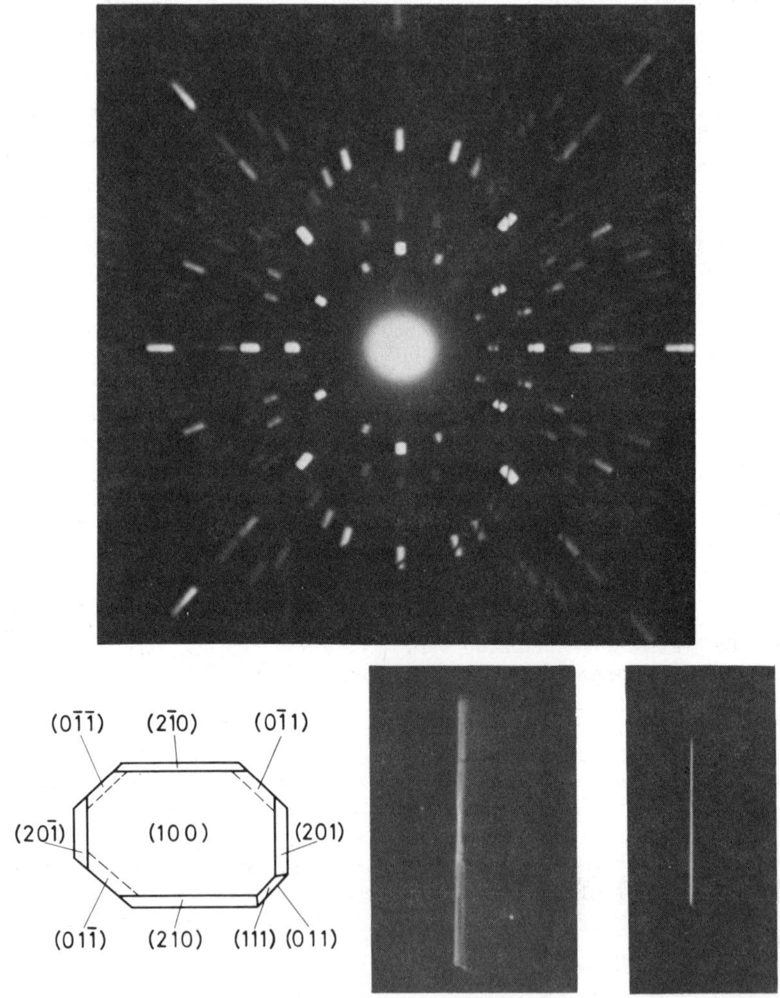

Figure 13b. x-Ray characterization of POM. Key: top, Laue diagram; bottom left, morphology; and bottom right, Lambot method compared to a similar sample of Si for which the penetration depth is smaller.

that of the binary axis Y. The angle between the binary axis Y and the displacement axis has been shown to have a constant value, 72°, for various samples grown from the melt in similar conditions. The Lambot method gives the degree of disorientation of molecules in the crystal. It shows the high degree of perfection obtained in the best POM monocrystals (see "figure 13"). Crystalline perfection of POM sample are presently further studied by means of X-rays topography. The absence of dipole-dipole interactions between POM molecules as well as between POM-acetonitrile molecules probably favors easier desorption of the solvent molecules at the interface seed-solution. The vanishing POM molecules dipole may thus contribute to the observed good chemical and structural quality of POM crystals as already conjectured when engineering the molecule.

Conclusion

We have described for a number of molecular single crystals the basic principles and techniques which were involved in the optimization of their nonlinear efficiency and crystalline quality. Certainly, other types of molecules or material processing techniques can be used. In particular, it could be worthwile to try and by-pass the difficulty of growing large single crystals for applications which are less demanding in terms of optical quality parametric oscillation.

However, developments such as follow, are expected, in the future to help both provide new insights in the nature of optical excitations in organics and help promote the use of organic materials in nonlinear devices :

- development of a "structural engineering" aiming at optimizing molecular orientations in the crystal which should ensure one order of magnitude increase of crystalline nonlinear susceptibility over MAP or POM (i.e. two orders of magnitude above $LiNbO_3$) with comparable molecular hyperpolarizabilities.
- research toward new processing methods of nonlinear organic materials (other than single crystals) : oriented organic thin films, inclusion of organics in polymers, organic glasses and others.
- extension of the molecular engineering approach which has proved successfull in the field of three-photon effects (χ^2) to four-photon effects (χ^3) with other potential applications : polymers, which are discussed at length in other chapters of this book, are very appealing for higher χ^3 values and currently studied by us (33) and others.

Acknowledgments

It is a pleasure to acknowledge fruitfull and continous assitance in various fields from R. Bertrand, D.S. Chemla, I. Ledoux,

H. Le Person, J.F. Nicoud, J.L. Oudar, A.M. Pougnet, J. Primot, M. Sigelle.

Literature Cited

1. Bloembergen, N. "Nonlinear Optics" ; Benjamin : New-York, 1965
2. Rabin, H. ; Tang, C.L. "Quantum Electronics" ; Academic : New-York, 1965
3. Oudar, J.L. ; Zyss, J. Phys. Rev.A 1982, 26, 2016
4. Zyss, J. ; Oudar, J.L. Phys. Rev.A 1982, 26, 2028
see also Oudar, J.L. ; Zyss, J. Polymer Preprints 1982, 23, 160
5. Davidov and al. J.E.T.P. Letters 1970, 12, 16
6. Jerphagnon, J. I.E.E.E. J. Quantum, Electron, 1971, 7, 42
7. Kurtz, S. ; Perry, T.T. J. Appl. Phys. 1968, 39, 3798
8. Murrell, J.N. "The Theory of the Electronic Spectra of organic molecules" ; John Wiley : New-York, 1963
9. Oudar, J.L. ; Chemla D.S. Opt. Commun. 1975, 13, 164
10. Zyss, J. ; Berthier, G. J. Chem. Phys. 1982, 77, 3635
11. Oudar, J.L. ; J. Chem. Phys. 1977, 67, 446
12. Hauchecorne, G. ; Kerherve, F. ; Mayer, G. Journal de physique 1971, 32, 47
13. Ledoux, I. ; Zyss, J. Chem. Phys. 1982, 73, 203
14. Oudar, J.L. ; Chemla, D.S. J. Chem. Phys. 1977, 66, 2664
15. Lalama, S.L. ; Garito, A.F. Phys. Rev. A 1979, 20, 1179
16. Morrell,J.A. ; Albrecht, A.C. Chem. Phys. Letters 1979, 64, 46
17. Zyss, J. ;Chem. D.S. Phys. 1979, 30, 3333 ; ibid, 1979, 70, 34I ibid. 1979, 71, 909
18. Zyss. J. ; Chemla, D.S. ; Nicoud, J.F. J. Chem. Phys. 1981, 74, 4800
Chemla D.S. ; Zyss J. ; Nicoud, J.F. ; French patent 7931455, 1979 (extended to USA UK and other countries)
19. Oudar, J.L. ; Hierle, R. J.Appl.Phys. 1977, 48, 2699
20. J.F. Nicoud ; Zyss, J. Coquillay. M. to be published
Nicoud, J.F. ; Zyss, J. French patent 8201166, 1982 (extended to USA, UK and other countries)
21. Flytzanis, C. French patent "Nonlinear Behaviour of molecules
Atoms and Ions in Electric, Magnetic or Electromagnetic 22.
Fields" ; Louis Neel, Ed ; Elsevier : Amsterdam 1979
23. Sigelle, M. ; Hierle, R. J. Appl. Phys., 1981, 52, 4199
Lipscomb, G.F. ; Garito, A.F. ; Narang, R.S. J. Chem. Phys.
24. 1981, 75, 1509
Knossow, M. ; Mauguen, Y. ; de Rango, C. Cryst. Struct.
25. Commun. 1976, 5, 723
26. Sangers Biochem. J. 1945, 39, 507
27. Fischer, E. Ber. Dtsch. Chem. Ges. 1901, 34, 433
28. Fischer, E. Ber. Dtsch. Chem. Ges. 1901, 34, 2868
29. Fischer, E. Ber. Dtsch. Chem. Ges. 1906, 39, 433

30. Frankel, M. ; Katchalski J. Am. Chem. Soc. 1942, 64, 2264
Gordon, M. ; Miller, J.G. ; Day, A.R. J. Am. Chem. Soc. 1948,
31. 70, 19
32. Brockmann, H. ; Musso, H. Ber. 1954, 87, 581
Rustow, B. ; Bohm, S. ; Ristau, O. Stud. Biophys. Berlin 1968
10, 2, 117
33. Kajzar, F. ; Messier, J ; Zyss J. ; Ledoux, I. Opt. Commun.
1983, 45, 2, 133

RECEIVED May 16, 1983

5

Characterization of Liquid Crystalline Polymers for Electro-optic Applications

G. R. MEREDITH, J. G. VANDUSEN, and D. J. WILLIAMS[1]

Xerox Corporation, Webster Research Center, Webster, NY 14580

In this paper we discuss properties of a novel medium displaying nonvanishing $\chi^{(2)}$. It is a medium whose properties are a synthesis of different classes of materials: liquid crystals, polymers and molecules possessing extremely large second-order nonlinear polarizability. We begin by briefly reviewing the special optics related properties of liquid crystals, polymers and liquid crystalline polymers (lc polymers), then show the synthesis of ideas which suggests pursuit of molecular doping of the latter. We discuss the physics of guest molecule alignment in a nematic host and the relationship of the response to DC poling fields and the induced $\chi^{(2)}$ tensors. Subsequently aspects of second harmonic generation (SHG) experimental characterization of $\chi^{(2)}$ are considered. Dynamical and temperature dependent aspects of SHG are presented which indicate that the lc polymer host plays more than a passive role in the poling-freezing process. We therefore present experimental investigations of associated properties of these lc polymers.

Liquid Crystals And Optics. In the fields of electro-optics and nonlinear optics, technologies develop as a consequence of the availability of specialized properties of certain materials. For example, the high degree of cooperative alignment of nonsymmetrically polarizable molecules in liquid crystals and the ease with which the direction of their alignments can be changed have lead to an inexpensive display technology (1). The ability to drastically alter the macroscopic optical properties with relatively low electrical fields have made liquid crystals attractive for other optical devices (e.g. light valves) as well. These effects are associated with DC or low frequency Kerr behavior (2). Also, because of the cooperativity, nonlinear optical phenomena such as the AC Kerr effect, self focusing and nonlinear refraction associated with phase conjugation are very pronounced in liquid crystal forming media, particularly in the vicinity of the liquid crystal - isotropic transition or Frederiks point (3,4).

[1]Current address: Eastman Kodak Research Laboratories, Rochester, NY 14650.

0097-6156/83/0233-0109$07.50/0

The unique properties of liquid crystals have also provided opportunity for study of novel nonlinear optical processes. An example involves the ability to modify the pitch of cholesteric liquid crystals. Because a pseudo-wave vector may be associated with the period of pitch, a number of interesting Umklapp type phase-matching processes (processes in which wave vector conservation is relaxed to allow the vector addition to equal some combination of the material pseudo-wave vectors rather than zero) are possible in these pseudo-one-dimensional media. Shen and coworkers have investigated these employing optical third harmonic generation (5) and four-wave-mixing (6).

Other than employing liquid crystals as media for interesting optical effects, linear and nonlinear optical experimentation can provide a probe into their structure as well. Spectroscopy of dichroism associated with light absorption by guest molecules dissolved in liquid crystal hosts has been used both to characterize the guest and to infer properties of the host (7,8). Spectroscopy and light scattering associated with the host is a more direct but perhaps more difficult probe of the latter. One interesting round of experimentation concerned optical second harmonic generation (SHG) and proof through its observation whether cholesteric media were acentric on the scale of molecular organization. In nematic media it is believed that a director defines the axis of axial alignment, there being no preferential polar alignment (9). Cholesterics can be viewed as having the same microscopic properties with the directors simply rotating on a macroscopic scale along the axis associated with pitch. In initial experiments probing this issue SHG was observed from cholesteric media (10) but was later suggested to have arisen as a consequence of the presence of noncentrosymmetric microcrystals (11). When care was taken to assure the exclusion of microcrystals no detectable harmonic light was produced (11,12). A decade later the very weak SHG associated with spatial dispersion in a nematic liquid crystal was reported (13). Recently it has been shown that if certain symmetry conditions are met, noncentrosymmetric liquid crystalline order on the scale of intermolecular organization can be achieved (14). The special characteristic required in smectic C phase to generate ferroelectric liquid crystalline media has been achieved by the inclusion of relatively small chiral centers on mesogenic molecules. Realizable macroscopic dipole polarization densities are consequently small. Use of this vehicle to achieve second-order nonlinearly polarizable media, as described below, is not expected to be rewarding. However, at least one electro-optic device using such materials has been devised (15). Another nonlinear optical phenomena, closely related to the main topic of this paper, which has been performed in liquid crystals is DC electric field induced second harmonic generation (16). In those experiments the DC field weakly modifies the equilibrium distribution of molecules to produce a net noncentrosymmetric polar alignment. SHG is subsequently observed. In these studies one may vary the polarizations of optical fields to observe the departure of anisotropies from those of isotropic media (vide infra). The variable pitch of cholesterics was also used to achieve pseudo-phase-matching of the type mentioned above for this process (17).

Polymers And Optics. The important special features of polymers in optical technologies are probably most strongly associated with their fabrication characteristics. Inexpensive lenses, prisms, fiber optics, antireflection coatings, etc. result from the ease with which plastic components may be produced. This is not to say that there have not been special polymers produced which have highly unusual or potentially beneficial optical properties. For example, the polydiacetylene materials, discussed in fuller detail in other articles of this volume, provide extended molecular orbital conjugation systems. The resulting delocalization of π-electrons can lead to very large third-order nonlinear polarizability and important optical properties associated with it (18). Special advantages of this system lie in the extremely rapid speed associated with purely electronic response to electric fields. Second-order nonlinearity has also been designed into this class of polymers (19). In this case, though, the second-order nonlinearity is associated with the pendant groups, probably being a consequence of local nonlinear polarizability. The polymeric property makes for a rugged medium in addition to offering the potential of patternwise (photo)polymerization and structure creation. An extra chore, though, in this approach is that one must identify a specific nonlinear monomeric species which crystallizes in a structure in which a significant fraction of the molecular nonlinear polarizability is preserved (a problem discussed in several papers in this volume) and grow suitable optical quality crystals before polymerization. The use of polymerization to rigidize structures is also beneficial when these species are drawn as thin film from Langmuir troughs, a versatile alignment tool (20,21). Other aspects of the theory of polymers in nonlinear optics may be found in the paper of Flytzanis in this volume.

A specific set of experiments which must be mentioned, being directly associated with the main topic of this paper, is the work of Bergman, et. al. (22) dealing with the second-order nonlinear optical properties of polyvinylidene fluoride (PVF_2). Nonvanishing $\chi^{(2)}$, the second-order nonlinear electric dipole susceptibility, is expected in PVF_2 since it exhibits other properties requiring noncentrosymmetric microscopic structure. These properties appear due to the crystallinity of PVF_2. $\chi^{(2)}$ associated with SHG was therefore measured in appropriately stress oriented and poled samples. Since the nonlinear polarizability in this case is associated only with saturated bonds, the magnitude of $\chi^{(2)}$ achieved (or achievable) is not significantly large ($\sim 2 \times 10^{-9}$ esu subsequent to poling in a 30V/μm field) in comparison to the nonlinearities discussed in this symposium. In this paper we will describe another polymeric structure, based on more recent knowledge of characteristics which enhance nonlinear polarizability, exhibiting larger $\chi^{(2)}$ and projected to display even higher values than we have reported (23).

In the study of molecular properties via optical spectroscopy, polymers have been extensively employed for the purpose of eliminating molecular rotation and inducing preferential orientation. Used as minimally perturbing matrices, polymers enable various spectroscopic polarization correlation studies. Uniaxial stress orienting of films is also a convenient means of achieving partial axial alignment of nonsymmetrical guest molecules. Recently Havinga and VanPelt (24) have taken another approach for spectroscopic study of large dye molecules dissolved in polymer matrices. They used large electric fields in conjunction with thermal modification of viscosity to orient and "freeze-in" net polar alignments of the guest molecules.

Liquid Crystalline Polymers. Recently it has been shown that if mesogenic species are attached to a polymeric backbone through sufficiently long and flexible spacer groups, such a medium will exhibit liquid crystalline behavior (25,26). The viscosity of such media is greatly increased over that of normal liquid crystals and switching times in liquid crystalline display devices are tremendously increased, although recent work has shown that these times can be made short. That goal is directly opposite to our goal of increasing rigidity for alignment preservation (vide infra). Also, an important consideration for some optical applications is that of beam attenuation by scattering. Associated with the increased response time for reorientation of the liquid crystal director is a decrease in the fluctuational light scattering which can be quite large in normal liquid crystals (27).

Molecularly Doped Thermotropic Liquid Crystalline Polymer. The idea of the nonlinear optical medium which is the subject of this paper results from a synthesis of the ideas of the discussion above and a few concepts from nonlinear optical molecular and crystal physics. As discusssed several places in this volume, it is known that certain classes of molecules exhibit tremendously enhanced second-order nonlinear polarizabilities, β. One of the main obstacles preventing the immediate technological use of these compounds is the orientational cancellation characteristic of polar third rank tensors. Only in noncentrosymmetric environments are their major components nonvanishing, and furthermore, only for highly correlated orientations of the microscopic species, do the β tensors constructively sum to yield a large value of $\chi^{(2)}$. Therefore, much research centers around the investigation of (noncentrosymmetric) molecular crystals as possible new nonlinear optical media which will possess the special advantages designed into the molecular constituents. Since many of the enhanced β species are nearly planar aromatics with large dipole moments, it's been speculated that minimization of the total crystal energy favors antiparallel molecular alignment to maximize the (negative) contributions of dispersion and dipole-dipole interactions. Obviously, the large dipole moments alternatively allow the possibility to achieve significant (partial) electric field induced polar molecular alignments. Following Havinga and VanPelt (24), one might use a very large static field to achieve an alignment of large β, large μ molecules dissolved at high concentration in a polymer matrix. By their method, reorientation is performed at high temperatures in a low viscosity state with subsequent cooling in the field to a higher viscosity state. The field is removed, leaving a noncentrosymmetric medium which they employed only for spectroscopic purposes. This medium should have nonvanishing $\chi^{(2)}$ if the polymer can, as they showed, permanently prevent relaxation of the alignment. The magnitude of $\chi^{(2)}$ would be determined by the degree of alignment, the concentration and the magnitude of β (and some local field phenomena).

We have proceeded one step further (23), expecting to use the inherent (axial) alignment properties of liquid crystals to enhance the net guest alignment achievable with an applied field. Besides this, use of polymers which are simultaneously liquid crystals increases the optical flexibility of the resultant medium since they are birefringent. It turns out, though, that the physics of our medium is more complex than this simple picture. The host does not act as a simple matrix with only the viscosity varying with temperature. Significant variation of the alignment is observed through SHG as function of temperature. Consequently, in this paper we will discuss our conception of the alignment physics and report characterizations of aspects of these liquid crystalline polymers relating to this problem.

Guest Alignment Physics

Nematic Potential. (28) For a guest molecule dissolved in liquid crystalline host, details of molecular shape and guest-host interactions determine the resultant orientational distribution. Of course, the guest introduction may modify host properties and guest-guest interactions may become important at the high concentrations we've used. Obviously, also, the guest alignment characteristics need not mimic those of the host. Since in these studies we have used a rather large rod-like guest molecule, 4−dimethylamino−4′−nitrostilbene (DANS), we expect it to associate nearly equally well with host mesogenic units or other guests (which is essentially verified by the relatively high concentrations achievable before crystallization occurs and by the properties reported below). Therefore, we adopt a mean-field description where the nematic potential, $U(\theta)$, determines the orientational distribution function, $P(\theta)$, of guest molecules in a nematic domain. $P(\theta)$ is the relative probability that a guest molecule will lie with its z axis oriented at angle θ to the nematic director (or laboratory reference axis). It is assumed, perhaps incorrectly for thin films, that P is independent of φ in spherical polar coordinates. Also it is assumed that the effects of rotational orientation of a molecule such as DANS about its "long axis" (direction for maximum interaction with the host), z, which is also herein assumed to be a good approximation to the direction of the permanent dipole moment μ, are averaged away.

In Figure 1 we show that for isotropic medium the renormalized distribution function,

$$P'(\theta)=P(\theta)/\sin\theta \qquad (1$$

is flat, i.e. there is no prefered orientational direction. On the other hand, the nematic potential causes a distribution which is peaked at $\theta=0$ and $\theta=\pi$ and which is symmetric about $\theta=\pi/2$ since $U(\theta)$ is. The relative values of $P'(\theta)$ are statistically determined by relative values of $\exp\{-U(\theta)/kT\}$. The nature of these distributions is commonly probed by experiments which determine specific moments. Perhaps the easiest to study is dichroism which indicates the deviation of the second moment from that of an isotropic distribution. The associated order parameter is

$$S^{(2)}=\langle(3\cos^2\theta-1)/2\rangle \qquad (2$$

which obviously is zero in isotropic and unity in totally parallel orientational distributions. Fourth moments can be determined from Raman (29) or fluorescence (30) depolarization among other

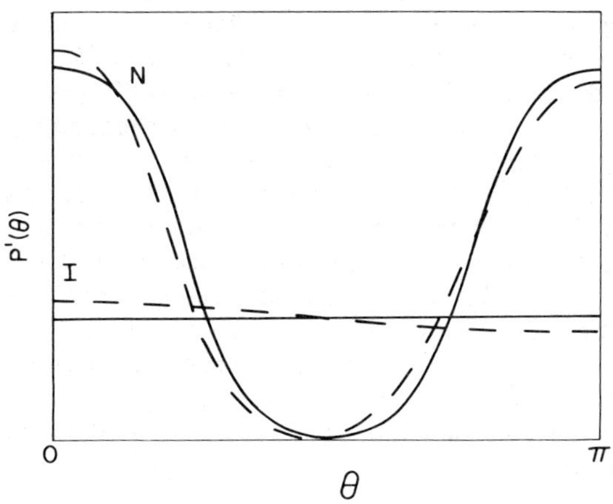

Figure 1. Schematic representation of statistical orientational distribution functions in isotropic (I) and nematic (N) potentials. Dashed curves depict adjusted distributions under electric field poling.

techniques. However, because of the uncertainty in electric local fields these optical techniques carry some unavoidable uncertainty, which is diminished if magnetic phenomena are used as probes of the orientational moments (9).

Poling Response. Application of a poling field in the isotropic case, even for very large fields and for dipoles of the magnitude of μ in DANS (7.4 D), only perturbs $P'(\theta)$ weakly since the $\underline{\mu}\cdot\underline{E}$ electrostatic energy is small compared to kT at or above 300 K. This behavior is represented in an exaggerated manner by the dashed curve I in Figure 1. A similar effect occurs in the liquid crystal when

$$U(\theta)'=U(\theta)-\underline{\mu}\cdot\underline{E} \quad (3$$

replaces $U(\theta)$ in determining $P'(\theta)$. This situation is also depicted by the exaggerated dashed curve N in Figure 1. (Obviously we must assume that perturbation of the liquid crystalline host distribution function is minimal. Otherwise one must express $U(\theta)$ as function of $\underline{E}$.) The net polar alignment $\langle\cos\theta\rangle$ can be calculated (31):

$$\langle\cos\theta\rangle = \int \cos\theta \exp\{-U'(\theta)/kT\} \sin\theta\, d\theta \Big/ \int \exp\{-U'(\theta)/kT\} \sin\theta\, d\theta \quad (4a$$

$$= \int \cos\theta \exp\{\mu E\cos\theta/kT\} P_o(\theta)\, d\theta \Big/ \int \exp\{\mu E\cos\theta/kT\} P_o(\theta)\, d\theta \quad (4b$$

$$= \int [\cos^2\theta\, (\mu E/kT) + \cos^4\theta\, (\mu E/kT)^2/6 + \ldots]\, P_o(\theta)\, d\theta \Big/ \int [1+\cos^2\theta\, (\mu E/kT)^2/2+\cos^4\theta\, (\mu E/kT)^4/24+\ldots]\, P_o(\theta)\, d\theta \quad (4c$$

Here we have used the zero-field nematic distribution function $P_o(\theta)$ for convenience of notation. The degree of net polar alignment can be seen to be enhanced in the liquid crystal over the isotropic case. The limiting cases are isotropic distributions and the Ising model (in which only $\theta=0$ and $\theta=\pi$ are allowed orientations). By retaining only the leading terms in the last equation one sees that in the high temperature limit

$$\langle\cos\theta\rangle_{Isotropic} = \mu E/3kT \quad (5a$$

$$\langle\cos\theta\rangle_{Ising} = \mu E/kT \quad (5b$$

These are the familiar orientational contributions to the DC dielectric response. This limit, $|\mu E/kT| \ll 1$, can be considered alternatively to be a restriction to nonsaturated alignments. For physical systems with orientational distributions intermediate between the isotropic and Ising limiting models the poling responses are easily related to $S^{(2)}$ since for $|\mu E/kT| \ll 1$ equation (4c) above states that

$$\langle \cos\theta \rangle = (\mu E/kT)\ \langle \cos^2\theta \rangle_{E=0} \qquad (6$$

Form of $\chi^{(2)}$. In our materials we are interested in properties of the induced $\chi^{(2)}$ tensor in a poled sample, that is, in behavior of the third moments <u><rrr></u>. These may be calculated in the same manner as $\langle Z \rangle = \langle \cos\theta \rangle$. Before proceeding with this task it is beneficial to make use of group theory and physical arguments to investigate the number of independent terms which must be considered or calculated. Since the poling field establishes $C_{\infty V}$ symmetry, the $\chi^{(2)}$ tensor may be shown to have the following form in contracted index notation (through tensor analysis it is known that in all C_{pV}: $p \geq 4$ point groups $\chi^{(2)}$ has the same form (32,33) and one may look-up the $\chi^{(2)}$ tensor form of C_{4V} (4mm) or C_{6V} (6mm) (34)):

	1	2	3	4	5	6
1	0	0	0	0	XXZ	0
2	0	0	0	XXZ	0	0
3	ZXX	ZXX	ZZZ	0	0	0

The sequential order of the ZXX, XZX and XXZ (laboratory reference frame) polarization labels is important only to the extent that frequency dispersion affects the magnitude of the elements of $\chi^{(2)}$.(34)

Considering dispersion, there are two importantly different types of nonlinearity which might be exhibited by a poled molecularly doped polymer. First, the partial polar alignment gives rise to a nonvanishing component of second-order nonlinear <u>electronic</u> polarizability which will allow purely optical processes such as SHG or optical wave mixing to occur. Second, in the electro-optic effects the effective refractive indices of the medium are altered by the application of a static electric field. If only the electronic nonlinearity makes a contribution, then not only will the so-called index interchange symmetry (i.e., the possible independence of magnitude of $\chi^{(2)}_{ijk}$ on the ijk sequential order referred to at the end of the last paragraph) be more likely to hold under nonresonant conditions, but $\chi^{(2)}_{SHG}$ will also be expected to be approximately equal to $\chi^{(2)}_{EO}$. However, it is possible that in the polymeric state where sub-T_g motion can occur the static field driven molecular rotation, enhanced by the large value of μ, coupled with the large anisotropy of α (the linear electronic polarizability) will provide a substantial response. This response is related to the DC Kerr effect which is an electro-optic effect (2). The latter is constrained to be second-order or higher in the strength of the DC field participating in the electro-optic effect by the usual complete vanishing of zero-field polar alignment in fluids. This would not be the case in the poled polymer system. In this paper we will concentrate on the first, totally electronic nonlinearity, but mention that we are pursuing evaluation of the electro-optic response.

Electronic $\chi^{(2)}$ And Its Molecular Source. The purely electronic $\chi^{(2)}$ can be related to the purely electronic β tensors of the molecules in the sample through a description of the manner in which their cancellation in the absence of electric field is negated by poling. Since $\mu \cdot \beta^{v}$ is very large in DANS (35), for description of our high concentration samples it is justifiable to neglect the contribution of the host lc polymer to $\chi^{(2)}$. (β^{v} is a portion of β which transforms as a vector in the three dimensional rotation group and is the only portion which may consequently contribute to poling induced $\chi^{(2)}$ in fluids(33).) Also since β_{zzz} is the only significant element of β in DANS (35), we only need to know the distribution of molecular z axes to approximately describe $\chi^{(2)}$. For this purpose, since only the two sets $\chi^{(2)}{}_{ZZZ}$ and $\chi^{(2)}{}_{P\{ZXX\}}$ are nonvanishing as shown above, one requires knowledge only of

$$\langle(\underline{k}\cdot\underline{K})(\underline{k}\cdot\underline{K})(\underline{k}\cdot\underline{K})\rangle = \langle\cos^3\theta\rangle \quad (7$$

and

$$\langle(\underline{k}\cdot\underline{K})(\underline{k}\cdot\underline{I})(\underline{k}\cdot\underline{I})\rangle = \langle(\cos\theta)(\sin^2\theta)(\cos^2\varphi)\rangle \quad (8$$

where $\underline{i}$, $\underline{j}$, $\underline{k}$ ($\underline{I}$, $\underline{J}$, $\underline{K}$) are the molecular (laboratory) cartesian unit vectors. Remembering the independence of a nematic system from the spherical polar coordinate φ,

$$\begin{aligned}\langle(\cos\theta)(\sin^2\theta)(\cos^2\varphi)\rangle &= \int(\cos\theta)(\sin^2\theta)P(\theta)d\theta\int(\cos^2\varphi)d\varphi \\ &= \int(\cos\theta-\cos^3\theta)P(\theta)d\theta/2 \\ &= (\langle\cos\theta\rangle-\langle\cos^3\theta\rangle)/2 \quad (9\end{aligned}$$

For isotropic media perturbed by the $-\mu\cdot\underline{E}$ electrostatic energy in the high temperature, or nonsaturated alignment ($|\mu E/kT| \ll 1$) limit, values are calculated to be

$$\langle\cos\theta\rangle_{Isotropic} = \mu E/3kT \qquad (10$$
$$\langle\cos^3\theta\rangle_{Isotropic} = \mu E/5kT \qquad (11$$
$$\langle(\cos\theta)(\sin^2\theta)(\cos^2\varphi)\rangle_{Isotropic} = \mu E/15kT \qquad (12$$

The associated poled frozen normal polymer susceptibilities are

$$(\chi^{(2)}{}_{ZZZ})_{Isotropic} = F\beta_{zzz}\mu E/5kT \qquad (13a$$
$$(\chi^{(2)}{}_{ZXX})_{Isotropic} = (\chi^{(2)}{}_{ZZZ})_{Isotropic}/3 \qquad (13b$$

where F is an unspecified local field correction factor. (We note that the last relationship relates to the requirement $\chi^{(3)}{}_{1111}=3\chi^{(3)}{}_{P\{1122\}}$ required by spatial and Kleinman symmetries in isotropic media which describe real-time third-order nonlinear optical effects, including the case when one of the fields is a static field such as in DC electric field induced second harmonic generation(36).)

In the Ising model similar results are

$$\langle\cos\theta\rangle_{Ising} = \mu E/kT \qquad (14$$
$$\langle\cos^3\theta\rangle_{Ising} = \mu E/kT \qquad (15$$

with the predictions

$$(\chi^{(2)}{}_{ZZZ})_{Ising} = F\beta_{zzz}\mu E/kT \qquad (16a$$
$$(\chi^{(2)}{}_{ZXX})_{Ising} = 0 \qquad (16b$$

Certainly for real samples where the distribution function is intermediate between the isotropic and Ising limits the susceptibilities also lie between. Two interesting points can be made. 1) Disregarding variation of F, the macroscopic nonlinearity can be enhanced by up to a factor of five over the maximum achievable in isotropic media by use of liquid crystal host; and 2) The nonlinearity which might be used for noncritical phase-matched SHG via the birefringence (which must depend on the magnitude of $\chi^{(2)}{}_{P\{ZXX\}}$) is diminished from that calculated for an isotropic distribution and approaches zero as the nematic distribution function becomes more tightly squeezed around $\theta=0$ and $\theta=\pi$. This is related to the fact that the projection of DANS z-axes onto the laboratory X-axis is reduced.

One may wish to pursue the bulk phase-matched behavior beyond the noncritical ($\delta=\pi/2$, δ defined below) case. In bulk samples birefringence is the usual means to overcome the natural dispersion which causes nonlinear polarization waves and harmonic light waves to travel with different phase velocities. (In thin film waveguides, one will have more flexibility which we do not discuss presently.) Consequently, either Type I or Type II methods (37) are employed to achieve equal velocities and maximum conversion. Since this medium is uniaxial, the implication is that the process will depend on either $\chi^{(2)}{}_{P\{oee\}}$ or $\chi^{(2)}{}_{P\{eoo\}}$. Here P indicates one of the set of possible permutations, and e and o indicate the extraordinary and ordinary light waves appropriate to propagation of light with wave vector tilted at an angle δ relative to the uniaxis (Z-axis). Taking the X-axis to lie in the plane of the Z-axis and the wave vector, the o-wave is polarized parallel to the Y-axis and disregarding the nontransverse character of the e-wave it is polarized parallel to $(\underline{I}\cos\delta+\underline{K}\sin\delta)$. It is easy to show, by substituting the appropriate polarization unit vectors as just described and refering back to the form of the $\chi^{(2)}$ tensor, that

$$\begin{aligned}\chi^{(2)}{}_{P\{oee\}} &= P\{\underline{o}\cdot\chi^{(2)}\cdot\cdot\underline{ee}\}\\ &= P\{\underline{J}\cdot\chi^{(2)}\cdot\cdot(\underline{I}\cos\delta+\underline{K}\sin\delta)(\underline{I}\cos\delta+\underline{K}\sin\delta)\}\\ &= 0\end{aligned} \quad (17a$$

and that

$$\begin{aligned}\chi^{(2)}{}_{P\{eoo\}} &= P\{\underline{e}\cdot\chi^{(2)}\cdot\cdot\underline{oo}\}\\ &= P\{(\underline{I}\cos\delta+\underline{K}\sin\delta)\cdot\chi^{(2)}\cdot\cdot\underline{JJ}\}\\ &= \chi^{(2)}{}_{P\{ZXX\}}\sin\delta\end{aligned} \quad (17b$$

Consequently only second-order processes employing one e- and two o-waves may occur and the second conclusion of the last paragraph applies beyond the noncritical phase-matched case.

To learn about the molecular behavior in the lc polymer, one might use measurements of $\chi^{(2)}{}_{ZZZ}$ and $\chi^{(2)}{}_{ZXX}$ as measures of $\langle\cos\theta\rangle$ and $\langle\cos^3\theta\rangle$. Since the latter are really asociated with the zero-field second and fourth moments, those measurements might be used with results of other optical techniques referred to earlier to evaluate consistency of the methods and local field models.

SHG In Thin Film Liquid Crystals

SHG, more so than other optical techniques, can give deceptive results due to propagation phenomena. Disregarding for the moment optical absorption effects, non-phase-matched SHG displays the behavior pictured in Figure 2 (38). The harmonic intensity $I^{2\omega}$ emerging from a sample depends on the sample thickness l in a

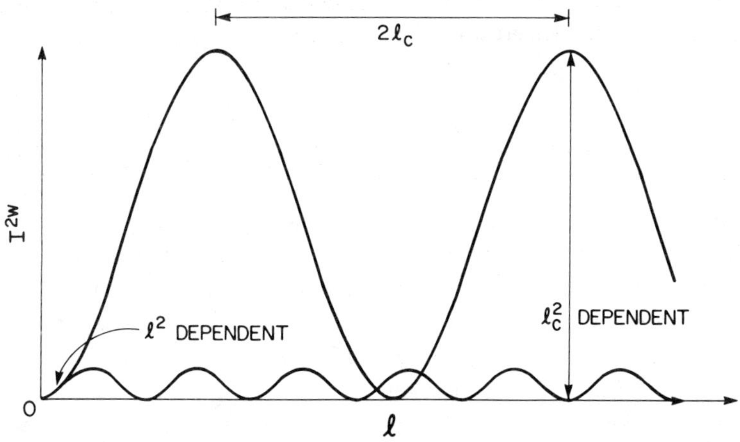

Figure 2. Depiction of the length dependence of second harmonic intensity. The lower curve shows the effect of decreasing the coherence length to 30% of the upper curve.

sinusoidally varying manner. The repeat distance is twice the coherence length, l_c. The amplitude of the sinusoidal behavior is directly proportional to $(l_c)^2$ so that care must be taken to determine 1) where one's sample thickness falls on such a curve and 2) what the value of l_c is before one can comment on the actual nonlinearity of the material. This point is illustrated by the appearance of a lower curve in Figure 2 which represents the relative behavior of the same material if its coherence length were reduced to 30% of the upper. However, for samples thinner than $\sim l_c/2$, $I^{2\omega}$ is independent of l_c and depends instead on $(l)^2$. This can be seen in the lower left hand corner of Figure 2 where the two curves coincide. Inclusion of weak linear absorption phenomena can be easily incorporated without changing this conclusion as we have described elsewhere (23).

Since for determination of $\chi^{(2)}_{ZZZ}$,

$$l_c = \lambda/4(n_Z^{2\omega} - n_Z^{\omega}) \tag{18}$$

where λ is the wavelength of the fundamental (laser), and since values of l_c for the 1064nm→532nm SHG we have used are typically 10—30 μm, films less than ~3 μm thick may be unambiguously used. For the material described herein we have done this by comparing $I^{2\omega}$ from a 3 μm thick film to $I^{2\omega}_{max}$ (the maximum signal obtainable from a sample by varying its thickness, that is, by translating a wedge across the laser beam) from an α—quartz crystal oriented for utilization of its $\chi^{(2)}_{XXX}$ component. $\chi^{(2)}_{XXX}$ in α—quartz has been accurately determined and is frequently used as secondary standard for SHG. Dealing appropriately with the weak attenuation at 532 nm as described elsewhere, we have determined $\chi^{(2)}_{ZZZ} \sim 3 \times 10^{-9}$ esu for a polymer with 2 weight percent DANS loading subjected to a 1.3 V/μm poling field (23).

To give some insight into the behavior of our sample, an Ising limit estimate of $\chi^{(2)}_{ZZZ}$ is useful. Based on the reported (35) DC field induced SHG of DANS in acetone solution $\langle\chi^{(2)}_{ZZZ}\rangle_{Ising} \sim 6 \times 10^{-9}$ esu. Discussion of the approximations, including local field treatment, is available in our earlier report (23). This value is a factor two larger than previously reported because of earlier neglect of the convention of reporting β in units analogous to the d coefficients of SHG, which are a factor 2 smaller than $\chi^{(2)}_{SHG}$, as commonly used (39). Since the ratio of observed and

Ising model estimate of $\chi^{(2)}_{ZZZ}$ reflects the ratio of alignment moments, $\langle\cos^3\theta\rangle_{DANS} \sim \frac{1}{2}\langle\cos^3\theta\rangle_{Ising}$. This reduction is essential for consistency with the observation of a dichroism order parameter $S^{(2)}=0.3$ or $\langle\cos^2\theta\rangle \simeq 0.5$. To further pursue the joint implications of these linear and second-order optical observations is perhaps unwise considering the uncertainties associated with solvent dependencies of the magnitude of β^V (40). Nevertheless, we have seen that a respectably large $\chi^{(2)}_{ZZZ}$ can be induced in these materials at relatively low poling fields. Since dielectric strengths in polymers are typically more than an order of magnitude higher than the poling field we employed, one can anticipate proportionately larger $\chi^{(2)}_{ZZZ}$ can be induced.

Turning to the investigation of the anisotropy of second-order nonlinearity, one must be cautious of evaluation of the magnitude of $\chi^{(2)}_{ZXX}$ performed as just described by simply rotating the plane of polarization of the laser beam. We tried this operation and saw a drastic decrease in $I^{2\omega}$. However, this may merely reflect the smaller magnitude of l_c in this SHG process. In this case

$$l_c=\lambda/4(n_Z^{2\omega}-n_X^{\omega}) \quad (19$$

Whereas dispersion determines $(n_Z^{2\omega}-n_Z^{\omega})$, which is ~0.01 in many transparent dielectric materials for 1064nm→532nm SHG, birefringence will dominate $(n_Z^{2\omega}-n_X^{\omega})$. Typically $(n_Z^{2\omega}-n_X^{2\omega})$ is 0.1–0.2 in liquid crystals. Coherence lengths in that case are in the range of 1¼–2½ μm. These lengths are smaller than our film thickness and one must be cautious of the propagation complexity described above.

Although we have not performed the evaluation to be described, we would like to point out a simple visual inspection which can remove most of the uncertainty in a comparison of harmonic intensities as just mentioned. The inspection relies on the dominance of the birefringence over dispersion and the similarity of SHG fringe behavior with birefringent waveplate transmission interference fringes. For a birefringent waveplate oriented with optical vibration directions tilted simultaneously 45° to the polarization directions of crossed polarizers between which it is placed, the spectral variation of the fraction of incident light passed by the analyzer can be usefully employed. The fraction of intensity passed relates directly to the relative phase retardation, $\Delta\varphi$, which has developed between the two orthogonal vibration modes of the waveplate. It is easy to show that $\Delta\varphi=2\pi\Delta n l/\lambda'$ where Δn is the difference in refractive indices of the two waves for light with vacuum wavelength λ'. Defining a wavelength dependent "passage" length

$$l_p = \lambda'/2\Delta n \qquad (20$$

it is also easy to show that the passed light intensity has the property $I_p \propto \sin^2(\pi l/2l_p)$. Noting that for SHG $I^{2\omega} \propto \sin^2(\pi l/2l_c)$ the similarity of phenomena is emphasized. Remembering the definition of l_c for this particular SHG process in equation (19), at $\lambda'=\lambda/2$ $l_c=l_p$ if dispersion is neglected. Consequently, if the film has some variation of thickness, as ours did, one need only observe the film between crossed polarizers and arrange to study 1064nm→532nm in the maximally green areas. Determination of the magnitude of $l_c \simeq l_p$ can be done by accurate thickness measurement and assessment of the order of this liquid crystal film waveplate. The latter could be achieved by either gradually thinning the sample to zero thickness (scraping or dissolution) or by observing the build up of fringes from the periphery of the sample where the aligning fields are nonuniform.

Dynamical and Thermal Dependences of Induced SHG

The physical description of DANS molecules aligning under the influence of the nematic potential and a poling field would seem to adequately describe the dichroism and $\chi^{(2)}$ measurements discussed above. However, special treatment of the sample had to be used to achieve those results (23). (Briefly, a film of DANS/lc polymer mixture was solvent cast over a 1000 μm spaced interdigital electrode pattern photolithographically delineated on a glass slide substrate. Aligned mesophase was prepared by application of approximately 500 V across the electrodes when the sample was heated near the clearing temperature. Poling to induce polar alignment, as will be shown shortly, had to be conducted at much lower temperatures.)

Laser Assisted Relaxation. We have poled a sample which was observed to retain alignment longer than a week (the longest time waited before reexamination). In contrast it was observed that excessive laser flux or insufficient thermal regulation of the films had major effect on the behavior of induced SHG. At higher laser powers local heating of the medium can occur. This is a reasonable expectation since DANS has a low lying electronic excitation which overlaps the SHG frequency (35). Either linear absorption of 2ω photons, or, more likely, direct nonlinear two-photon (or higher) excitation can deposit excess energy directly at the DANS molecules. The energy of

2ω—photons or two ω—photons is ~90kT at 300 K. If this absorbed energy flows in the short term into n modes of the system, this energy can in a transient manner raise the local temperature by a factor of [(90+n)/n]. Also, though we have no evidence for this, changes in local structure due to the modified properties of the higher electronic DANS state can occur. By applying, removing and reversing polarity of the poling field and monitoring the time dependence of the SHG intensity one may then probe the local relaxation and reorientation behavior. We described such a study in our earlier report and reproduce the time dependent observations of Figure 3.

Prior to time zero of Figure 3 the sample had been poled for several hours at 1.3 V/μm (which is the magnitude of the poling field applied subsequently and pictured at top of the figure). The SHG showed somewhat complex response to variation of the applied field. We attribute the rapid SHG buildup and decay components immediately following changes in the poling field to rotation of the DANS molecule inside a "cage". The "cage" is determined by the polymer and hinders the complete loss of DANS alignment. Consequently its much slower relaxation is evidenced by the slower, weak relaxation components of SHG. Also, the reduced response to reversed polarity poling can be attributed to the incomplete relaxation of these cages. They are appropriate to alignment in the original polarity, having been created by the continued aligning forces in that direction over several hours, and consequently hinder realignment of DANS molecules in the opposite direction. (We note that this is a slight deviation from the nematic potential model. In Figure 1 the poled distribution, as actually drawn, reflects this distortional shift of the host. For the constant nematic potential model, the poled distribution would be totally above or below the unpoled distribution for θ less than or greater than $\pi/2$, respectively.) Further, the increase of SHG on repeated reapplication of the original polarity field is taken as evidence of the non-smooth nature of the polymeric host; repeated orientational relaxation and realignment provides increased opportunity for DANS molecules to circumvent high local barriers to maximum rotation in the direction of the field. These barriers may be associated with DANS motion relative to "permanent" host structures or may be associated with slight reorganization of the host configuration as the DANS molecule drags along the mesogenic groups with which it must be strongly associated. These motions are not drastic since one may estimate the average angular rotation of DANS molecules to be ~1° (the net alignment being ~1% of saturation).

Considering the small rotations involved with the effects we have observed and considering the thermodynamic instability of the poled configuration, one wonders about the factors which stabilize the alignment. Two possibilities are impurity charge or strain compensation of the internal polarization fields. It is possible, then, that the explanation of laser assisted relaxation of SHG is really more complex than the simple model proposed above. It is also possible that part of the effect is associated with reconfiguration of local complexes driven by transient (pulse length ~20 nsec., pulse-to-pulse period=100msec. in our experiments) temperature changes correlated with the explanation we offer below for the drastic thermal dependence of induced SHG. The difference between the simple molecule-in-a-cage explanation and this being the expectation to find specific association complexes which are destabilized by the locally delivered laser energy.

Temperature Dependence of Induced SHG. We also reported earlier that the induced $\chi^{(2)}$ was strongly temperature dependent. Figure 4 shows the equilibrium temperature dependence of induced SHG in these materials. These are considered equilibrium properties since the same points were achieved, with sufficient time allowed for establishing equilibrium, when approached from below or above in temperature. The dashed 1/T curve is included to contrast the observed behavior with the high temperature, small alignment prediction of the model discussed above (DANS molecules responding to a poling field in a temperature independent nematic potential). Several speculative explanations of this behavior may be offered. An important accompanying observation is that the temperature region in which the drastic alteration occurs is the region of a broad second-order transition observed in DSC characteristic of polymeric T_g.

One explanation for the behavior depicted in Figure 4, which is easily dismissed, is that the change is due to drastic temperature dependence of $U(\theta)$. The maximum effect (Ising→Isotropic) which could arise is a change in intensity by a factor of ~25 (remembering that $I^{2\omega} \propto (\chi^{(2)})^2$), a smaller effect than observed. Also the strong dichroism was not visually seen to change in this temperature region. Alternatively, we suggest that since DANS is a molecule expected to form association complexes with itself and with the mesogenic groups, the competitive equilibria among DANS containing species is altered as the nature of the host changes through this temperature range. It should be emphasized that only association species which are noncentrosymmetric in their DANS content 1) can be significantly aligned in a polar sense through the resultant species dipole moment and 2) can contribute significantly through non-cancellation of β to the $\chi^{(2)}$ of SHG. This is similar to the well known phenomena that in permanent dipole measurements in liquids associations lead to dipole correlation factors, g, substantially differing from unity. In the temperature range in which various host motions are becoming

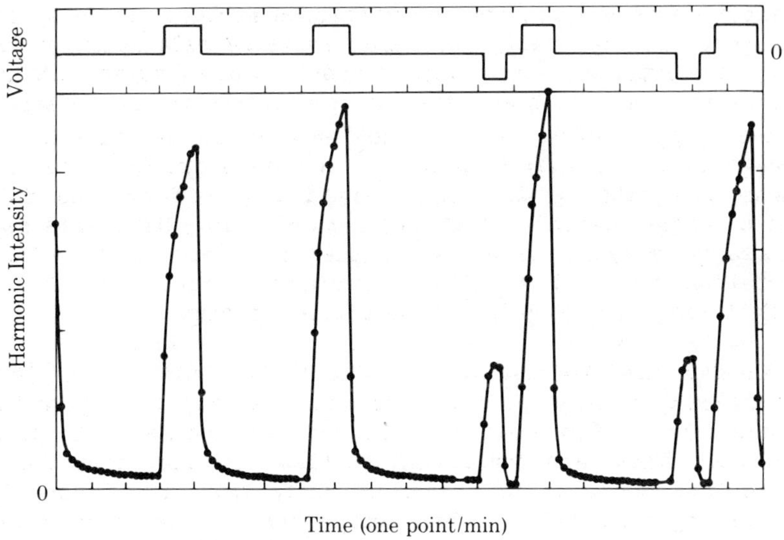

Figure 3. Time dependence of harmonic generation (of polymer/guest induced polar alignment and relaxation) when the laser intensity is overly strong. The poling-field time dependence is depicted at top. (Reproduced from Ref. 23. Copyright 1983, American Chemical Society.)

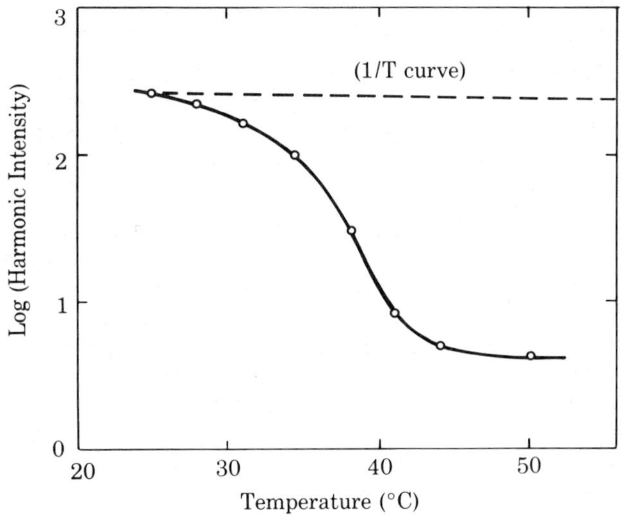

Figure 4. Temperature dependence of the induced second harmonic intensity. (Reproduced from Ref. 23. Copyright 1983, American Chemical Society.)

thermally actuated, one might expect to see some of these association constants or correlation factors to be substantially affected.

The $\chi^{(2)}$ inducing and alignment relaxation behaviors of this medium are strongly affected by host characteristics and those properties are associated with the T_g-like transition observed in DSC. We have, therefore, undertaken to synthesize lc polymers with shorter flexible spacer groups with the goal to increase the temperature of this transition. In the remainder of this paper we report synthesis, and thermal and dielectric relaxation characterizations.

Properties of the Liquid Crystalline Host Polymer

The lc polymers which we have studied are special cases of the general type pictured in Figure 5. The magnitudes of n and m must be such that decoupling of the mesogenic groups from the polymeric backbone is sufficient for liquid crystalline cooperative orientation to occur. It should be commented that copolymers are required for our nonlinear applications since y=1 polymers while displaying liquid crystal phase are not field orientable and while inclusion of the cyano species creates this property the x=1 polymers are amorphous. The material whose DANS doped properties were described above was that described by x=y=0.5 and m=n=6. Synthesis of monomers, structural identification, polymerization, molecular weight determinations, film preparation and optical characterization techniques have been described elsewhere (23,25,41).

Several other liquid crystalline variations of this class of polymers were produced for study. These had variously x, y=0, 0.5, 1 and m, n=3, 6. Figure 6 displays the thermal properties of these polymers observed by DSC. All exhibited two transitions as summarized in Table 1. The higher temperature transition (identified with T_m) correlates with the observation of loss of "texture" observed in polarizing microscopy associated with the loss of birefringence at the liquid crystalline phase (probably nematic) to isotropic melt phase transition (clearing temperature, T_c). The lower temperature transition has been associated with the glass transition exhibited by polymers. The degree of decoupling of the mesogenic groups from the backbone is strongly affected by change of n and m as evidenced by the progression in T_g. The weaker dependence of T_m may reflect the fact that motion in the backbone has already been thermally activated and this transition is associated with loss of correlation among the similar mesogens.

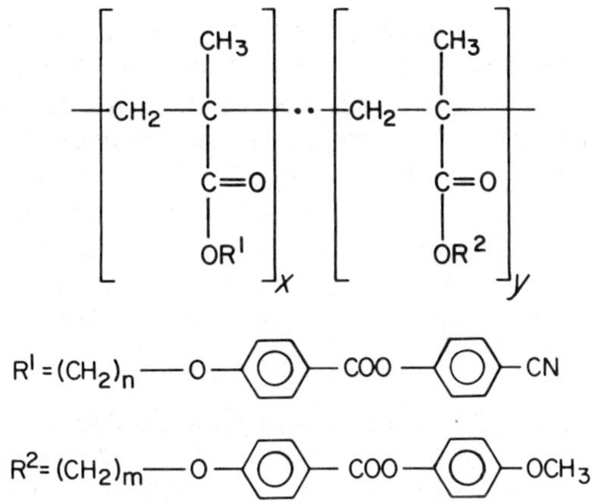

Figure 5. Liquid crystalline polymer family of this study.

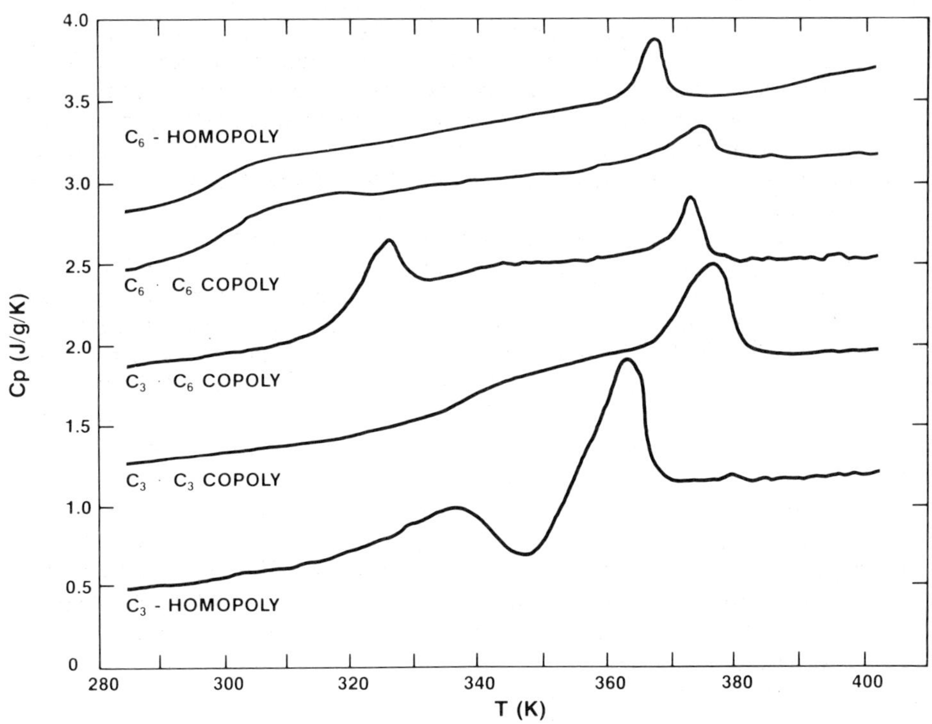

Figure 6. Differential scanning calorimetry scans of 1c polymers. Curves are displaced vertically. Conditions: heating, 20 K/min; and cooling, 320 K/min.

Table 1. Properties of lc Polymers

Sample	Spacer	M_w	T_g(DSC)	T_m(DSC)	T_c(Optical)
Homopolymer (y=1)	$-C_6-$	47K	27°C	95°C	95–96°C
Copolymer	$-C_6+C_6-$	33K	25°C	101°C	101.5°C
Copolymer	$-C_3+C_6-$	37K	44°C	101°C	102°C
Copolymer	$-C_3+C_3-$	–	57°C	103°C	105°C
Homopolymer (y=1)	$-C_3-$	46K	44–47°C	92°C	99°C

In order to increase the stability of the field induced alignment of DANS it may be necessary to increase the rigidity of the alignment medium. Decreasing n and m raises T_g presumably because shorter spacer groups increase the coupling of the interacting pendant groups to the backbone with a resulting increase in backbone rigidity. Conversely, it is difficult to predict if an increase in T_g resulting from this approach will adversely affect the alignability of the mesogens or the loss of polar DANS alignability above T_g.

The loss of alignment of the doped species above T_g is contrary to the situation described by Havinga and VanPelt (24) where the decreased internal viscosity above T_g is accessed to aid in the poling procedure. Since we have observed maximum polar alignment to be achieved at or below T_g, it is difficult to predict the effect of a more rigid medium. Obviously, some freedom of motion is required for this alignment to occur below T_g. Figure 7, which is the temperature dependent dielectric loss behavior at 1 KHz, shows that indeed there are sub–T_g motions active in these polymers at room temperature. It is apparently the fortunate weak coupling of such motions to the DANS alignment motions which both allows the slow (several hours) poling to proceed and allows the alignment stabilization factors to operate before relaxation can occur.

Conclusion

A novel second-order nonlinear optical medium which should offer considerable fabrication flexibility has been described. The physics of alignment of the highly nonlinearly polarizable moiety was discussed. However, observation of complex dynamical and thermal behavior indicates that an important role is played by the polymer liquid crystalline host. Additional properties of modified members of this family of lc polymers were consequently investigated. The explanations of guest alignment stabilization and thermal dependence of the alignability remain unresolved issues.

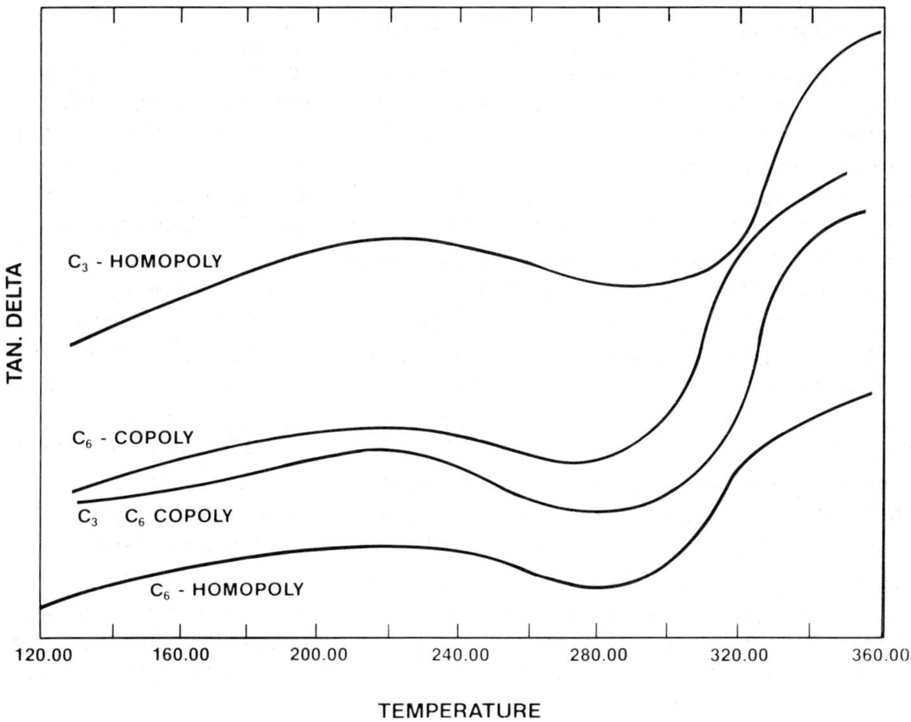

Figure 7. Temperature dependence of dielectric loss of lc polymers. Curves are displaced vertically.

Acknowledgment

We thank J. Pochan, F. Roberts, R. LaDonna, R. Hudson, W. Herbert and N. Nowacki for experimental contributions.

Literature Cited.

1. Margerum, J. D.; Miller, L. J. J. Colloid Interface Sci. 1977, 58, 559.
2. Hellwarth, R. W. Prog. Quantum Electron. 1977, 5, 1.
3. Khoo, I. C. Phys. Rev. 1982, A25, 1040 and 1637.
4. Wong, G. K. L.; Shen, Y. R. Phys. Rev. 1974, A10, 1277.
5. Shen, Y. R. "Nonlinear Spectroscopy"; Bloembergen, N., Ed.; North-Holland: New York, 1977, p. 201.
6. Ye, Peixuan; Shen, Y. R. Appl. Phys. 1981, 25, 49.
7. Levenson, R. A.; Gray, H. B.;Ceasar, G. P. J. Am. Chem. Soc. 1970, 92, 3653.
8. Cox, R. J. Mol. Cryst. Liq. Cryst. 1979, 55, 1.
9. Stephen, M. J.; Straley, J. P. Rev. Mod. Phys. 1974, 46, 617.
10. Freund, I.; Rentzepis, P. M. Phys. Rev. Lett. 1967, 18, 393.
11. Durand, G.; Lee, C. H. Mol. Cryst. 1968, 5, 171.
12. Goldburg, L. S.; Schnur, J. M. Rad. Electron. Eng. 1970, 39, 279.
13. Arakelyan, S. M.; Lyakhov, G. A.; Chilingaryan, Yu. S. Sov. Phys. Usp. 1980, 23, 245.
14. Meyer, R. B. Mol. Cryst. Liq. Cryst. 1977, 40, 33.
15. Clark, N. A.;Lagerwall, S. T. Appl. Phys. Lett. 1980, 36, 899.
16. Saha, S. K.;Wong, G. K. Appl. Phys. Lett. 1979, 34, 423.
17. Saha, S. K. Opt. Comm. 1981, 37, 373.
18. Sauteret, C.; Hermann, J. P.; Frey, R.; Pradere, F.; Ducuing, J.; Baughman, R. H.; Chance, R. Phys. Rev. Lett. 1976, 36, 956.
19. Garito, A. F.; Singer, K. D.; Hayes, K; Lipscomb, G. F.; Lalama, S. J.; Desai, K. N. J. Am. Opt. Soc.1980, 70, 1399.
20. Tieke, B.; Graf, H. J.; Wegner, G.; Naegele, B.; Ringsdorf, H.; Bauerjie, A.; Day, D.; Lando, J. B. Colloid Poly. Sci. 1977, 255, 36.
21. Pitt, C. W.; Walpita, L. M. Thin Solid Films 1980, 68, 101.
22. Bergman, J. G.; McFee, J. H.; Crane, G. R. Appl. Phys. Lett. 1971, 18, 203.
23. Meredith, G. R.; VanDusen, J. G.; Williams, D. J. Macromol. 1982, 15, 1385.
24. Havinga, E. E.; VanPelt, P. Ber. Bundsenges. Phys. Chem. 1979, 83, 816.
25. Finkelman, H.; Ringsdorf, H.; Wendorf, J. H. Makromol. Chem. 1978, 179, 273.
26. Shibaev, V. P.; Kostromin, S. G.; Plate, N. A. Eur. Polym. J. 1982, 18, 651.

27. Whinnery, J. R.; Hu, Chenming; Kwon, Y. S. IEEE J. Quantum Electron. 1977, 13, 262.
28. Wojtowicz, P. J. "Introduction to Liquid Crystals"; Priestley, E. B.; Wojtowicz, P. J.; Sheng, Ping, Eds.; Plenum: New York, 1975.
29. Jen, S.; Clark, N. A.; Pershan, P. S.; Priestley, E. B. Phys. Rev. Lett. 1973, 31, 1552.
30. Penchev, I.; Dozov, I.; Kirov, N.; Afanasyeva, N.; Ruoliene, J. Spec. Lett. 1982, 15, 265.
31. Böttcher, C. J. F. "Theory of Electric Polarization"; Elsevier: New York, 1952, p. 28f.
32. Meredith, G. R. J. Chem. Phys. 1982, 77, 5863.
33. Jerphagnon, J.; Chemla, D.; Bonneville, R. Adv. Phys. 1978, 27, 609.
34. Butcher, P. N. "Nonlinear Optical Phenomena"; Ohio State University Engineering: Columbus, 1965.
35. Oudar, J. L. J. Chem. Phys. 1977, 69, 446.
36. Flytzanis, C. "Quantum Electronics: A Treatise"; Rabin, H.; Tang, C. L., Eds.; Academic: New York, 1975, Vol. I, p. 9.
37. Kurtz, S. K. "Quantum Electronics: A Treatise"; Rabin, H.; Tang, C. L., Eds.; Academic: New York, 1975, Vol. I, p. 227.
38. Byer, R. L. "Nonlinear Optics"; Harper, P. G.; Wherrett, B. S., Eds.; Academic: New York, 1977, p. 47.
39. Zernike, F.; Midwinter, J. E. "Applied Nonlinear Optics"; Wiley: New York, 1973.
40. Singer, K. D.; Garito, A. F. J. Chem. Phys. 1981, 75, 3572.
41. VanDusen, J. G.; Williams, D. J.; Meredith, G. R. Polymer Preprints 1982, 23, 149.

RECEIVED June 14, 1983

6

Second-Order Nonlinear Media from Spiropyran Merocyanine Quasicrystals

G. R. MEREDITH and D. J. WILLIAMS[1]

Xerox Corporation, Webster Research Center, Webster, NY 14580

S. N. FISHMAN, E. S. GOLDBURT, and V. A. KRONGAUZ

The Weizmann Institute of Science, Department of Structural Chemistry, Rehovot, Israel

Photoinduced, spontaneous aggregation processes have been shown to occur when indolinobenzopyrans are irradiated in aliphatic solvents. The aggregates which are globular in appearance, consist of submicron cores of crystalline materials with an amorphous exterior and are termed "quasicrystals" (1-3). Spectroscopic studies by Krongauz and coworkers (1) indicate that the composition of the cores are A_nB (n=2,3) and the amorphous exteriors AB. The most stable quasicrystals have been derived from 1-(β-methacrolyloxyethyl)-3,3 dimethyl-6'-nitrospiro-(indoline-2,2'-[2H-1] benzopyran (SP-A) and its associated merocyanine form (SP-B).

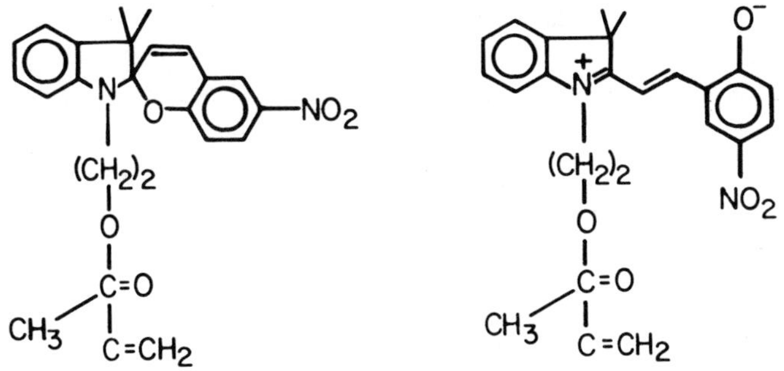

SP-A SP-B

When the irradiation of SP-A in hydrocarbon solvents is performed in an external electric field, thread-like structures are formed

[1]Current address: Eastman Kodak Research Laboratories, Rochester, NY 14650.

0097-6156/83/0233-0135$06.00/0

between the electrodes. The threads tend to be composed of globules of only one size and the globule sizes increase in a stepwise fashion. A model has been suggested (2) where the globule cores are actually formed from a particle consisting of associated J-aggregate type stacks with a diameter of approximately 40 nm. Higher order globules contain discrete numbers of individual particles (Table I).

Table I
Size and core parameters of globules of different generations

Generation	Average globule diameter (nm)	Particles in core
1	80	1
2	140	6
3	200	36

The large molecular hyperpolarizability of the merocyanine chromophore (4,5) and the highly polar environment of the quasicrystals has prompted studies of the second order nonlinear optical properties of these materials(6).

The photoinduced processes leading to the isomerization and complexation which precedes spontaneous aggregation have been the subject of considerable recent interest (7-10). By monitoring the light intensity and concentration dependences of the species produced following the photoexcitation of SP-A by picosecond and nanosecond, transient spectroscopy, a mechanism was developed which accounts for formation of the species preceding quasicrystal formation in the picosecond to nanosecond time domain (10). The schematic diagram below summarizes that mechanism. A and B indicate the ring closed and isomerized forms. ${}^1A^*$ is the first excited singlet state, ${}^3A^*$ the lowest excited triplet state, X is a ring opened form with the C-O bond broken but where the geometry of the parent compound is retained. The intermediacy of various short lived states such as X^* (a vibrationally hot form of X) and ${}^3A^{*\dagger}$ (an $n\pi^*$ or vibrationally excited $\pi\pi^*$ triplet) have been proposed as intermediates in

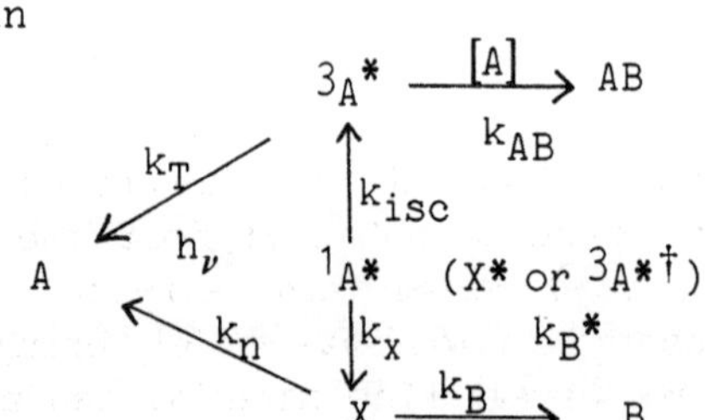

the formation of ${}^3A^*$, X, and B(9). The rate constants have the following values.

$$k_{isc} \sim k_x < 1.2 \times 10^{11}\ s^{-1}$$

$$k_B > k_n \sim 4 \times 10^9\ s^{-1}$$

$$k_T < 10^5\ s^{-1}$$

$$k_{AB} \sim 10^{10} M^{-1} s^{-1}$$

From the mechanism and values of the rate constants, the formation of B occurs very rapidly; within a few hundred picoseconds and AB is formed on the microsecond time scale. These species exhibit characteristic absorption bands in the 550 to 600 nm region of the spectrum. At very long times, i.e. several seconds of steady state irradiation, the red shift in the absorption band is complete and presumably due to A_nB as suggested by Krongauz (1-3). Thus far, it has not been possible to clearly time resolve the formation of aggregates from AB dimers, although subtleties in the transient absorption indicate this is occurring. For instance, the time resolved buildup in absorbance at the red end of 600 nm band seems to be slower than it is 10 or 20 nm further to the blue. This may indicate a process such as:

$$nAB \rightarrow (AB)_n$$

or

$$AB + A \rightarrow A_2B$$

$$nA_2B \rightarrow (A_2B)_n$$

are occurring which lead to globule formation.

In this paper we will briefly review the findings of an earlier study of the properties of quasicrystals using SHG and present further experimental data to probe their structure and nonlinear optical properties. Finally, a model describing the behavior of quasicrystals in an external electric field is developed.

Experimental Methods

Compound SP-A was prepared according to the method of Zajtseva, et al., (11). The final product obtained from the synthesis was purified on a silicic acid column and recrystallized from 1:10 hexane. Positive identification was made through IR and NMR analysis. The colorless crystalline material had a melting point which was heating rate dependent and had a maximum value of 105°C.

Globules and quasicrystals were prepared as described previously (2). In general, a solution of SP-A in methyl-cyclohexane is placed between brass electrodes with 1mm spacing physically contacted with a glass cover plate and a field of 1.5 kV/mm is generated between the electrodes. Irradiation of the solution results in thread-like structures which remain on the glass slide when the electrodes are removed and the solvent allowed to evaporate. A micrograph of threads deposited from a 2.5×10^{-3} M solution of SP-A in methyl-cyclohexane (MCH) is shown in Figure 1. The deposit shown in the figure is considerably more dense than that used for nonlinear optical characterization because of the high concentration of SP in the solution. Electron microscopic examination of a similar sample clearly illustrates the globular structure and thread-like morphology with highly anisotropic ordering resulting from deposition in the field (Figure 2). The relative content of globules of various generations was estimated from examination of several micrographs and found to consist of, 20% of threads containing first generation, 70% second generation and 10% third generation. Globules formed in the absence of field are shown in Figure 3. By lowering the concentration of SP-A in MCH and controlling the irradiation flux, it was possible to generate material which was composed of at least 90% first generation quasicrystals. As will be discussed below, the interpretation of the SHG experiments is facilitated by restricting the samples to primarily first generation globules.

For SHG experiments, samples deposited on a coverslip were sandwiched with an additional coverslip. Prior to creating the sandwich structure, thin strips of Al foil were placed on the spots originally occupied by the electrodes and a thin layer of petroleum jelly was applied to the threads for refractive index "matching". The Al strips, which protruded from the sandwich structure, were later connected to a high voltage source. A non-blocking configuration for the electrodes was required to prevent artifacts due to field collapse during application of external field.

The characteristics of the laser system and SHG apparatus have been described earlier. The fundamental wavelength was obtained from a Q switched Nd^{+3}/YAG laser operating at 1.064 μm and the harmonic was detected by a photomultiplier tube monochrometer combination with spectral filtering at 532 nm.

Results

The SHG characterization of the quasicrystals was conducted in two stages. The first study was conducted on samples with morphologies similar to those illustrated in Figure 2 where a substantial

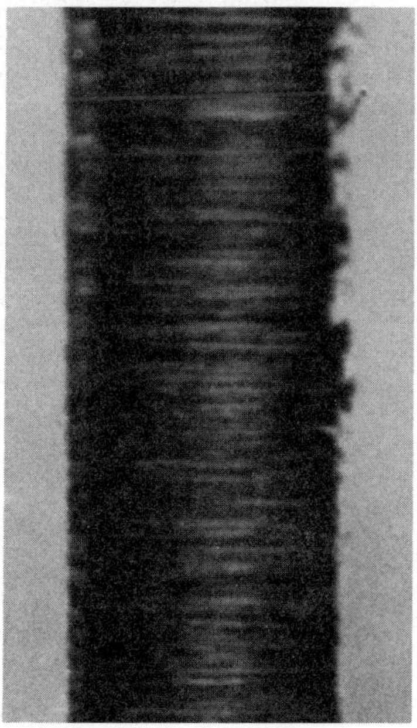

Figure 1. Optical micrograph of quasicrystalline threads. (Reproduced with permission from Ref. 6. Copyright 1982, North Holland Publishing Company.)

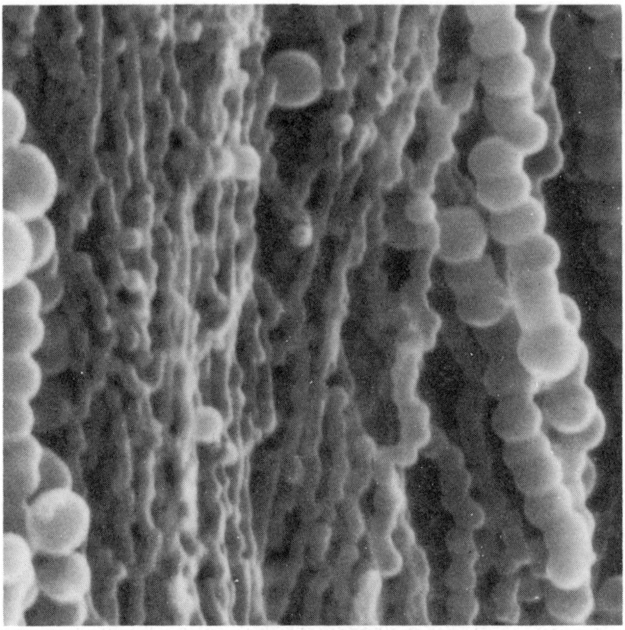

Figure 2. Scanning electron micrograph of globules formed in an electric field at 1.5 kV/mm. (Reproduced with permission from Ref. 6. Copyright 1982, North Holland Publishing Company.)

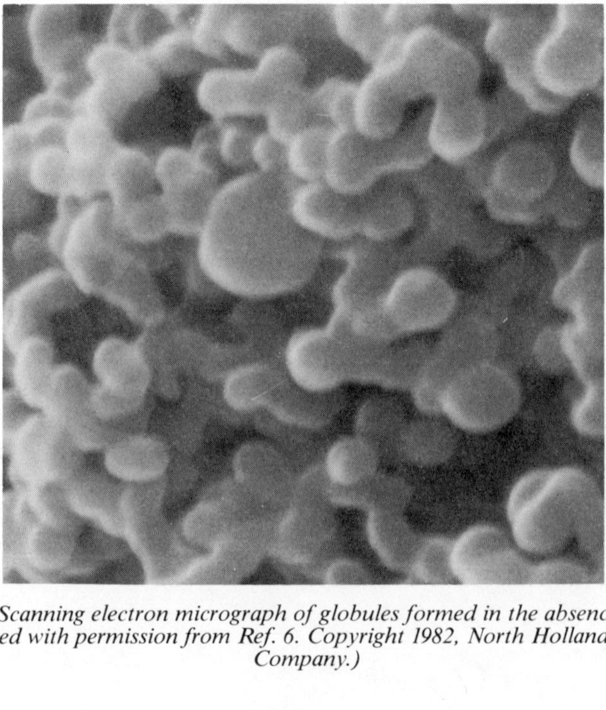

Figure 3. Scanning electron micrograph of globules formed in the absence of a field. (Reproduced with permission from Ref. 6. Copyright 1982, North Holland Publishing Company.)

fraction of the threads contained second or higher generation globules. These experiments were discussed previously (6) and are briefly reviewed here.

An important feature of the SHG study is its sensitivity to noncentrosymmetric ordering. Because of this, many unusual features concerning the behavior of quasicrystals in electric fields were obtained that were not manifest in any other way. Consistent with other indicators of noncentrosymmetry quasicrystals formed in the presence of external fields exhibit SHG. The SHG intensity and its optical plane of polarization were shown to be sensitive to the direction of the applied field in which they were formed, giving signal an order of magnitude less when the fundamental polarization was perpendicular to the thread direction than when it was parallel. The harmonic fields were themselves equally well polarized. It was also found that when external fields were applied, changes in the harmonic intensity and polarization could be induced. Comparison of the magnitude of the SHG signals with that obtained from nitrobenzene by the electric field induced second harmonic generation technique (EFISH) established that the effect was due to a true second order phenomenon and not simply a static third order response involving the electric fields used for perturbation. Furthermore, no SHG was observed from globules produced in the absence of the aligning field, but SHG could be induced in them by subsequent electric field application. Memory and field restoring tendencies associated with the field dependence of SHG were also seen.

We also found that the tendency of the SHG signal to respond to field changed with time. The response time gradually decreased and the assymetry in response to field direction was gradually lost. Optical and electron microscopic examination of the threads subjected to perturbing fields showed no discernable change in their morphology. Since the samples contained threads consisting largely of second or higher generation globules, it was thought that the field induced changes in SHG subsequent to thread formation may be due to rearrangement or relative reorientation of individual dipolar particles within the cores. Internal damage to the globule structure, particularly the amorphous component, may have occurred when the field disturbances were prolonged. Since these observations did not provide a clear explanation of the behavior of quasicrystals in relatively low electric fields, an additional set of experiments were conducted on threads composed primarily of first generation threads where the complexities of relative reorientation of particles within the core could be eliminated.

The freshly prepared sample exhibited SHG prior to application of an external field, confirming that the as formed threads are non-centrosymmetric. Figure 4 shows a plot of SHG intensity versus applied electric field under various conditions. The point indicated by START is the SHG intensity of a freshly prepared sample which had not yet been subjected to additional external fields. Following the arrows, one can trace the field treatment and resultant behavior of the sample. One observation to be emphasized is associated with equilibration. If the field is very slowly increased to +2000V, the SHG intensity is given by point 1. When the field is suddenly switched off, the signal returns to point 2 which is larger than its value at START and slowly decays to its original value over a period of several minutes. When the field is suddenly switched on, the signal assumes the value at 3 and over a much longer period, returns to point 1. Similar fast and slow responses are observed at other fields.

Switching the polarity of the voltage clearly indicated an assymetry in the behavior of the SHG signal. The equilibrated signal at -2000V (point 4) is approximately half the value of 3. The curve defined by the points taken after apparent equilibration (several minutes) at each field approximates a parabola with its origin at -450V.

Figure 5 shows a logarithmic plot of the steady state value of the SHG intensity versus applied voltage. The nearly quadratic response expected for a parabola is evident. A 20% variation from quadratic behavior would result in points within the limits indicated in Figure 5.

Discussion

The measurements of SHG and its electric field dependence extends our understanding of the nature of crystalline order in the quasicrystals. Originally the conclusion that the globule cores were crystalline was based on x-ray diffraction, electron diffraction from individual globules, and the dichroism associated with threads of globules formed in electric fields. A strong red shift in the optical absorption band suggested J-aggregate (Scheibe stack) structures within the cores. It was suggested earlier that a parallel arrangement of molecular dipoles within the stacks is energetically favored when the angle between the molecule and stack axis is less than 54° and the antiparallel configuration favored when it is greater than 54°. The large permanent dipole moment of the stacks and red shift are consistent with the parallel arrangement

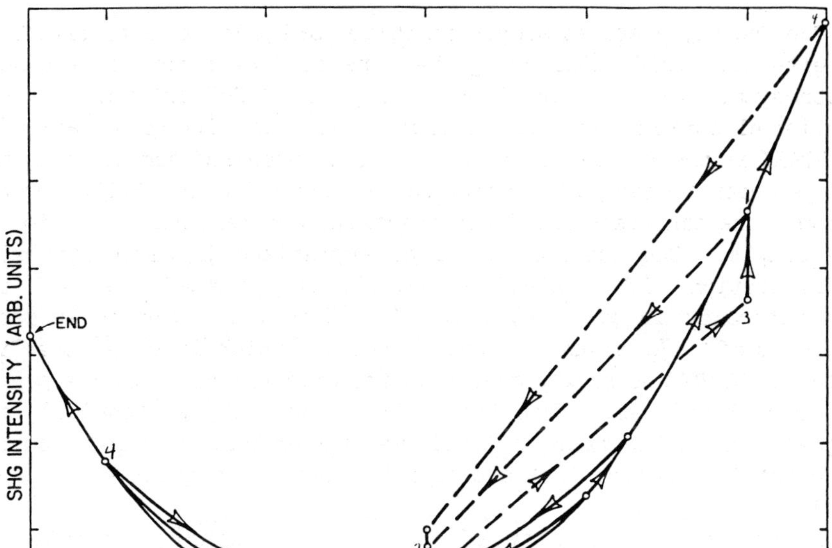

Figure 4. Plot of SHG intensity vs. electric field under various conditions. See text for explanation. (Reproduced from Ref. 13. Copyright 1983, American Chemical Society.)

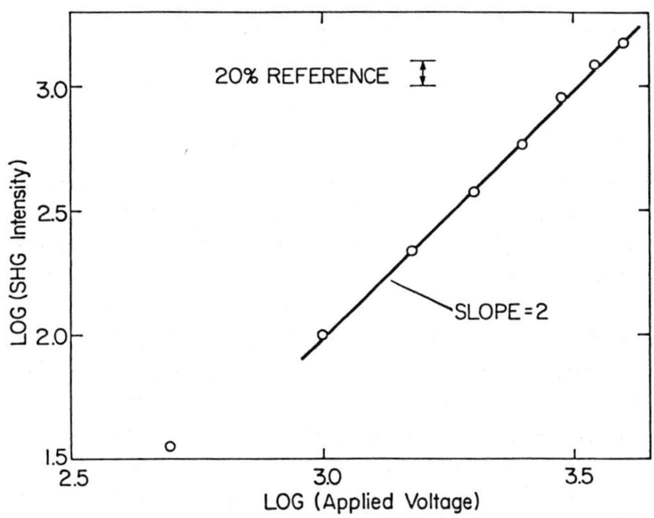

Figure 5. Plot of log SHG intensity vs. log applied field (20% uncertainty would produce the indicated variation). (Reproduced from Ref. 13. Copyright 1983, American Chemical Society.)

(2). These conclusions are examined in the light of the SHG measurements and a model is developed to explain the overall behavior below.

Before discussing the model based on SHG results, some additional comments concerning the nature of the SHG experiment are necessary. Since SHG is a coherent process, the phase relationships among all the oscillating polarization densities in the sample and all phase shifts occurring as the harmonic field propogates through the sample to the point of observation are important. On a molecular level, the relationship between the phases of oscillating polarization can be used to understand why centrosymmetric media composed of noncentrosymmetric molecules results in no observable frequency doubling. Inversion of a molecule with respect to a nearby neighbor causes a π phase shift in the component of polarization oscillating at frequency 2ω, which is proportional to the square of the fundamental field. This pair of molecules would broadcast effectively no harmonic field if the two are separated by a distance much less than the wavelength of light. In homogeneous media such as solutions or amorphous solids or centrosymmetric crystals, the cancellation is essentially complete (with the harmonic molecular polarizations being most easily detected through hyper-Rayleigh scattering).

When dealing with small noncentrosymmetric particles whose spacing is near but less than the wavelength of light, the question of mutual orientation and spacing becomes crucial. Consider a pair of noncentrosymmetric particles related by inversion symmetry and separated by 1/10 the wavelength of light. Their spacing is sufficiently large that they would broadcast regions of constructive and destructive interference in a quadrupole pattern. In the nodal regions there would be little harmonic intensity and in the lobes the interplay between the magnitude of the nonlinear response of each of the oscillators and the phase retardation to a particular observation point in space determines the magnitude of the harmonic field. In our experiment, we monitor the radiation intercepted by an F/16 cone normal to the plane containing the threads. That plane is also normal to the laser beam so that to the extent that the wavefront is planar and the sample optically thin, all the globules experience approximately the same phase retarded fundamental fields. In the example we would expect to see a limited amount of harmonic, even though the individual particles might show an extremely large nonlinear response.

As was indicated earlier, the particle sizes that we have produced by irradiation are of the order of 1/10 the wavelength of light.

Therefore, before any conclusions can be drawn from a specific experiment the details of globule orientation and placement must be considered.

One of the explanations we have considered for the electric field dependence of SHG (6) is that the individual single crystalline particles are rotated within the core, thus increasing the net alignment in the field direction. In this study, an effort was made to restrict the cores to first generation material and the behavior was very similar to that observed for cores with higher generation multiple particle cores. From the considerations developed above, it is not clear that rotation of individual particles with the geometrical constraints of higher generation quasicrystals should have a significant impact on the observed SHG. Secondly, we would have expected the first generation particles to retain their alignment due to the same dipolar interaction between globules within a thread which is considered to be responsible for the unique one dimensional thread formation. It is also difficult to understand how rotation of particles within the amorphous envelope can be accomplished without destroying the globules.

An alternative explanation for the observed behavior is that the external electrostatic field brings about reorientation of individual dipoles within a stack resulting in an increase in stack dipole moment. We will consider the implications of such a model in considerable detail.

Consider a stack of individual dipoles as illustrated in Figure 6 in which the dipoles are arbitrarily parallel or antiparallel to each other and at an angle α to the stack axis. The number of parallel and antiparallel neighboring pairs are determined by equilibrium conditions. If this Ising model is restricted to nearest neighbor interactions, analytic expressions for the alignment and electric field behavior can be obtained. Taking the interaction energies of neighboring dipolar pairs to be $u(\uparrow\uparrow)=u(\downarrow\downarrow)$ and $u(\uparrow\downarrow)=u(\downarrow\uparrow)$ and taking the energies of the individual dipolar species to be $u(\uparrow)$ and $u(\downarrow)$ independent of the orientation of nearest neighbors, define the following thermodynamic quantities

$$s = \exp\{-u(\uparrow\uparrow)/kT\} \qquad (1)$$

$$\zeta\sigma = \exp\{-u(\uparrow)/kT\} = \zeta \exp\{\mathbf{p}_{\uparrow}\cdot\mathrm{E}/kT\} \qquad (2)$$

ζ is a factor taking the internal energy into account and σ does so for the energy of the molecular dipole $\mathbf{p}_{\uparrow}$ in an applied electric field E. One can choose a reference potential such that

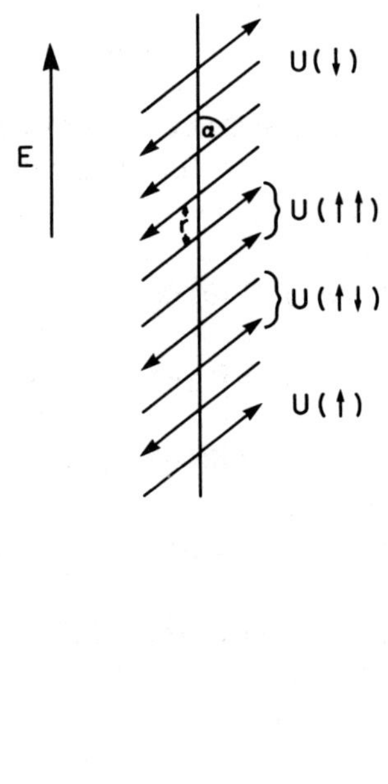

Figure 6. Schematic diagram of the model molecular stack with interaction energies U(↑↑), U(↑↓), and single site potentials U(↑), U(↓) indicated.

$$s^{-1} = \exp\{-u(\uparrow\downarrow)/kt\} \quad (3)$$

Also since $\mathbf{p}_{\downarrow} = -\mathbf{p}_{\uparrow}$,

$$\exp\{-u(\downarrow)/kT\} = \zeta\sigma^{-1} \quad (4)$$

Using the matrix method of calculation (14), it is possible to show that the classical partition function of the system is:

$$Z = \mathrm{Tr}\,(\mathbf{G}^n) \quad (5)$$

where $\mathbf{G}$ is the 2 x 2 matrix

$$\mathbf{G} = \begin{bmatrix} \sigma s & \sigma^{-1}s^{-1} \\ \sigma s^{-1} & \sigma^{-1}s \end{bmatrix} \quad (6)$$

The probability distribution functions of pairs and singletons were calculated from Z according to theoretical expressions and taking into account Eqs. 5 and 6:

$$w\uparrow\downarrow = n^{-1}[\partial \ln Z/\partial \ln(\sigma^{-1})] = (r^2-q^2)/2r(1+r-q) = w\downarrow\uparrow \quad (7)$$

$$w\uparrow\uparrow = n^{-1}[\partial \ln Z/\partial \ln(\sigma s)] = (r+q)/2r(1+r-q) \quad (8)$$

$$w\downarrow\downarrow = n^{-1}[\partial \ln Z/\partial \ln(\sigma^{-1}s)] = (r-q)/2r(1+r-q) \quad (9)$$

$$w\uparrow = w\uparrow\uparrow + w\uparrow\downarrow = \{1+(\sigma-\sigma^{-1})[(\sigma-\sigma^{-1})^2+4s^{-4}]^{-1/2}\}/2 \quad (10)$$

$$w\downarrow = w\downarrow\downarrow + w\uparrow\downarrow = \{1-(\sigma-\sigma^{-1})[(\sigma-\sigma^{-1})^2+4s^{-4}]^{-1/2}\}/2 \quad (11)$$

where

$$q = (\sigma-\sigma^{-1})/2\sigma \quad (12)$$

$$r = [q^2 + \sigma^{-2}\,s^{-4}]^{1/2} \quad (13)$$

In the absence of a static electric field (i.e., $\sigma=1$, $q=0$, and $r=s^{-2}$) s alone determines the correlations among neighboring dipoles.

$$w\uparrow\downarrow = w\downarrow\uparrow = 1/2(1 + s^2) \tag{14}$$

$$w\uparrow\uparrow = w\downarrow\downarrow = 1/2(1 + s^{-2}) \tag{15}$$

$$w\uparrow\uparrow/w\uparrow\downarrow = s^2 \tag{16}$$

A value for s much less than unity implies that antiparallel ordering is favored. Large values for s favors parallel alignment with antiparallel pairs appearing as imperfections between regions of parallel alignment. The red shifted J-aggregate type spectra suggests the parallel alignment, large s case is applicable to our system.

From equations (10), (11), (14), (15) and the definitions for s, the net alignment Q, is given by:

$$Q = (w\uparrow - w\downarrow) = \sinh\{p_\uparrow \cdot E/kT\}/[\sinh^2\{p_\uparrow \cdot E/kT\} + s^{-4}]^{1/2} \tag{17}$$

The influence of the electric field E and magnitude of nearest neighbor interactions on the total alignment can be determined by examining $\partial Q/\partial E$, and $\partial^2 Q/\partial E \partial s$ under various limits. It can be shown that $\partial Q/\partial E$ has one sign over the range of E implying that Q, $w\uparrow$, $w\downarrow$ are monatonic functions of E. It can be seen that nearest neighbor interactions significantly influence the degree of alignment induced by the field. Larger s improves the electric field alignment relative to the noninteracting case (s=1) provided certain limits are not exceeded. These limits are associated with the saturation of the alignment and are determined from the derivative $\partial^2 Q/\partial s \partial E$. For E satisfying $\cosh\{2p_\uparrow \cdot E/kT\} < (1+s^{-4})$ increased neighbor interactions increase the electric field alignment susceptability, but at larger fields saturation will again occur. An illustrative case of the effects of s and E occurs near E=0 where the net alignment is given by the first term in the expansion of (17)

$$(Q)_{E=0} \simeq s^2 (p_\uparrow \cdot E/kT) \tag{18}$$

Here the sensitivity to s is evident.

For comparison with experiment, several terms in the expansion are examined

$$Q = s^2X - (6)^{-1}(3s^6-s^2)X^3+(120)^{-1}(45s^{10}-30s^6+s^2)X^5+\ldots \quad (19)$$

where $X=p_{\uparrow}\cdot E/kt$. As indicated earlier, s is greater than unity for the molecular stackequationn the quasicrystals so that equation (19) converges quickly only for small X. The results in Figure 2 showed a quadratic dependence of SHG on E. Since SHG intensity is proportional to Q^2, only the leading term contributes within our ability to determine deviations from quadratic behavior. Taking the ratio of the first to second terms and assuming that deviations of 20% from quadratic behavior could be detected, an upper limit on s can be determined:

$$s_{max} = \{[1 + 0.6/(x_{max})^2]/3\}^{1/4} \quad (20)$$

At this point, it is instructive to emphasize that s_{max} is a statistical factor associated with the maximum interaction energy $u(\uparrow\uparrow)_{max}$ and $u(\downarrow\uparrow)_{max}$ consistent with the quadratic dependence of SHG on E, within the limits of experimental certainty. From s_{max} and estimates of the dipole moment, it is possible to calculate the limits on $u(\uparrow\uparrow)$ and $u(\downarrow\uparrow)$ using equations (20), (1) and (3).

$$[u(\uparrow\downarrow)-u(\uparrow\uparrow)] \leq L = (kT/2)\ln\{[1+0.6(kTp_{\uparrow}\cdot E_{max})^2]/3\} \quad (21)$$

Although we do not have an exact value for the molecular dipole moment, we will consider the implications of dipole moments in the range of 2 to 30D. Taking $\alpha \sim 35°$ as a noncritical estimate of the angle of the dipole with respect to the polar axis, values of L for a series of dipole moment values are shown in Columns 1 and 2 of Table II. In the range of dipoles moments of interest, the limit on L is ~0.1 eV. These limits are comparable to the heats of fusion of molecular crystals which makes the limit quite reasonable.

Table II. Values of Interaction Energies for Various Molecular Dipole Moments

$P\uparrow$(D)	L(eV)	$[u(\uparrow\downarrow)-u(\uparrow\uparrow)]_{d-d}$(eV) for $\alpha=0$	α_{min} for $[u(\uparrow\downarrow)-u(\uparrow\uparrow)]_{d-d}\sim 0.1$ eV
2	0.13	0.025	0°
5	0.11	0.015	25°
10	0.09	0.6	49°
20	0.07	2.5	53°37'
30	0.06	5.6	54°18'

An interesting comparison can be made between the experimentally determined limits on the interaction energies and those obtained from a simple dipolar model. In this case, the interaction energy difference between two coplanar dipoles is given by:

$$[u(\uparrow\downarrow)-u(\uparrow\downarrow)]_{d-d} = -(|\mathbf{p}_{\uparrow}|^2/|r|^3)2(1-3\cos^2\alpha) \qquad (22)$$

where r, $\mathbf{p}_{\uparrow}$, and α are defined in Figure 4. Assuming the parallel dipole configuration is the more stable according to arguments given above, restricts $\alpha < 54°44'$. Assuming r=7.4Å, one of the interplanar spacings observed from x-ray diffraction patterns (16), the maximum values of $[u(\uparrow\downarrow)-u(\uparrow\uparrow)]_{d-d}$ occurring at α=0 are given in Table II, Column 3. In the range of dipole moments of interest, these values are an order of magnitude too large. By assuming the maximum value of $[u(\uparrow\downarrow)-u(\uparrow\downarrow)]_{d-d}$ is ~0.1 eV minimum values for α for various dipole moments can be calculated and these are given in Column 4 of the table. Dipole moments in excess of 10D, require values of α very near the 54° value where the dipolar interaction energy is minimized. From the values in Columns 3 and 4, several points become apparent. The first is that if dipole moments of the molecules were in the 20 to 30D range, other intermolecular interactions (17) would have to stabilize the α ~ 54° configuration in order to maintain pair-wise interactions in the range of 0.1 eV consistent with the experiment and model. Because of the enormous dipolar energy, the system would want to favor much small α unless these mitigating interactions exceeded 2 eV. This situation hardly seems realistic. A second possibility is that the molecules are large polarizable π systems and the dipole-dipole model exaggerates the value of the interaction. Restricting the value of the intermolecular potential to that predicted by the dipole-dipole interaction is also unrealistic. Quantum mechanical (17) effects (e.g., dispersion forces) and quadrupolar effects due to the proximity of the delocalized charge distributions could significantly alter the potential.

Conclusions

Quasicrystalline threads composed of predominately first generation globules which were formed in an electric field exhibit a strong SHG signal, verifying their noncentro-symmetric character. Application of an additional external field was found to enhance the SHG signal in an asymmetric manner with roughly a quadratic dependence on the applied field. A model based on interacting stacks of nearest neighbor molecular dipoles was proposed and analytical solutions for the alignment of molecules in the stack as a function of molecular dipole moment, and nearest neighbor pair-wise interaction energy, were obtained. The solutions predicted the observed quadratic field behavior and the experimental uncertainty in the limits of the agreement was used to establish an upper limit on the intermolecular interaction energy. An independent estimate was made of the interaction energy by calculating the energy of an array of interacting molecules in the dipole-dipole approximation. Self consistency with the experimental data and model is obtained if the molecular dipole moments are assumed to be less than 10D and/or other aspects of the intermolecular interaction reduce the effective dipole moment.

Obviously, the model is crude and does not take into account many of the factors operating in a real molecular stack. Lack of symmetry with respect to the polar axis and the fact that dipoles may not necessarily be situated in one plane represent additional complications. The angle α could also be field dependent which is ignored in the model. The model also requires that interactions between molecules in adjacent stacks be very weak in order for fields of 10 to 20KV/cm to overcome barriers for field induced reorientation. The cores are then presumably composed of a more or less ordered assembly of stacks with a structure similar to smectic liquid crystals.

Several important aspects of the SHG experiments are not described in a straight forward way by the model. These are the residual SHG prior to field perturbation and asymmetric response to fields of different polarity. These effects may be due to the fact that the dipoles within the stacks as formed are subjected to remnant fields from surrounding stacks. The asymmetry may be associated with structural asymmetry within the stacks or some higher ordering or arrangement which does not allow for a symmetric hysteresis about zero voltage. A distribution of nonidentical stacks is also possible.

Literature Cited

1. Krongauz, V.A.; Parshutkin, A.A. Photochem. Photobiol. 1972, 15, 503.
2. Krongauz, V.A.; Fishman, S.N.; Goldburt, E.S. J. Phys. Chem. 1978, 82, 2469.

3. Krongauz, V.A. Israel J. Chem. 1979, 18, 304.
4. Dulcic, A.; Flytzanis, C. Opt. Comm. 1978, 25, 402.
5. Levine, B.F.; Bethea, C.G.; Wasserman, E.; Leenders. L. J. Chem. Phys. 1978, 68, 5042.
6. Meredith, G.R.; Krongauz, V.A.; Williams, D.J. Chem. Phys. Lett. 1982, 87, 289.
7. Krongauz, V.A.; Kiwi, J.; Gratzel, M. J. Photochem. 1980, 13, 89.
8. Kalisky, Y.; Williams, D.J. Chem. Phys. Lett. 1982, 86, 100.
9. Krysanov, S.A.; Alfimov, M.V. Chem. Phys. Lett. 1982, 91, 77.
10. Kalisky, Y.; Orlowski, T.E.; Williams, D.J. J. Phys. Chem. submitted.
11. Zajtseva, E.L.; Prokhota, L.H.; Kurkovskoya, L.H.; Shifrina, R.R.; Kardash, N.S.; Drapkina, D.A.; Krongauz, V.A. Kim. Giterotsikl. Soedin. 1973, 10, 1362.
12. Meredith, G.R. Rev. Sci. Instrum. 1982, 53, 48.
13. Meredith, G.R.; Williams, D.J.; Fishman, S.N.; Goldburt, E.S.; Krongauz, V.A. J. Phys. Chem., in press.
14. Kramers, H.; Wannier, G. Phys. Rev. 1941, 60, 252.
15. McClellan, A.L. Table of Experimental Dipole Moments, Rohera Enterprises, El Cerrito, 1974, 2, 633.
16. Krongauz, V.A.; Goldburt, E.S. Nature 1978, 271, 43.
17. Nordlund, K.; Ames, A.; Taylor, T. Photo Sci. Eng. 1970, 14, 295.

RECEIVED June 6, 1983

7

Nonlinear Optical Interactions in Organic Crystal Cored Fibers

B. K. NAYAR

British Telecom Research Laboratories, Martlesham Heath, Ipswich, Suffolk, IP5 7RE, United Kingdom

A novel technique of producing waveguiding structures by growth of crystals in glass capillaries is presented. This method of crystal growth is simple and is particularly suited to organic materials. The phase-matching techniques for non-linear interactions in these fibres are discussed and optical second harmonic generation is demonstrated in monomode benzil crystal cored fibres.

Non-linear optical interactions occur in materials with high optical intensities and have been used to produce coherent light over a wide range of frequencies from the far infra-red to the ultra-violet. The three wave mixing process is of particular interest as it can be used for optical parametric amplification and optical second harmonic generation (SHG) and occurs in non-centrosymmetric materials.

In recent years a number of organic materials having second order non-linearity of the same order and greater than that of commonly used inorganic materials have been reported (1-4). These crystals are mainly those having molecules which include benzene rings or conjugated bonds with acceptor or donor groups. Another advantage in using these materials is that, unlike some inorganic materials, they exhibit resistance to optical damage at high optical intensities. To use these materials for optical mixing it is necessary to phase match the interacting waves as materials generally exhibit chromatic dispersion. For some materials it is possible to exploit the material birefringence for phase matching and this method is known as 'Angle Phase-Matching'. However, for materials which are isotropic or do not possess adequate birefringence, it is possible to correct phase mismatch using quasi-phase-matching techniques (5-6). These in practice tend to be difficult to implement and offer low conversion

0097-6156/83/0233-0153$06.00/0

efficiencies. Alternatively, if they can be used to fabricate a waveguiding structure, phase matching can be achieved using waveguide dispersion.

In this paper, the advantages of non-linear interactions in waveguiding structures are discussed and we report on the fabrication of organic crystal cored optical fibres. These fibres have a crystalline core and a glass cladding. This method of waveguide fabrication is simple and is particularly suited to organic materials. The optical SHG was recently demonstrated (7) in benzil crystal cored fibres and here further results are presented and methods of improving SHG efficiency are discussed.

Non-linear Interactions in Waveguides

The advent of integrated optics has highlighted advantages of using waveguiding structures for non-linear interactions and these have been exploited by a number of researchers for optical SHG and mixing. These advantages are:

1. Phase matching can be achieved using waveguide dispersion. The waveguide dispersion arises because the phase velocity of a light wave of a given wavelength in a waveguide is determined by the guiding region dimensions, refractive index difference between the guiding region and the substrate, and the mode of propagation. In practice the phase matching is achieved for the interacting modes by suitably tailoring dimensions of the guiding region and/or the refractive index difference.

2. The dimensions of the guiding region for optical wavelengths are of the order of micrometers and as a result it is possible to have large optical intensities, required for non-linear interactions, with modest optical powers. Also, as the optical energy is guided it is possible to have long interaction lengths. An approximate comparison of the efficiency of SHG process in a bulk medium and that in a waveguiding structure can be made by considering the product of the intensity of the beam at the fundamental frequency and the interaction length.

For bulk material:

$$I_\omega L=(P/\pi w_o^2)(\pi w_o^2/\lambda)=P/\lambda$$

where, gaussian beam diffraction has been assumed and w_o is the beam radius at the focus.

Whereas, for a three dimensional guide:

$$I_\omega L=(P/\lambda^2)L=(P/\lambda)(L/\lambda)$$

where, it has been assumed that the guide dimensions are comparable to the wavelength of the fundamental wave.

The advantages described above of using waveguiding structures can cancel out if the field overlap integral of the interacting modes is small. This arises since the SHG efficiency is proportional to it.

$$\eta_{SH} \propto \int \underline{E}_{\omega} \cdot \underline{E}_{2\omega} \, dA$$

where,

$\underline{E}_{\omega}$ is the electric field distribution of the fundamental mode.
$\underline{E}_{2\omega}$ is the electric field distribution of the SH mode.
A is the waveguide cross-sectional area.

The overlap integral places constraints on mode types into which SH can be coupled and is maximum when both the fundamental and the SH propagate as fundamental waveguide modes. In practice, phase matching is generally achieved for SH as a higher order mode due to the waveguide dispersion characteristic and this leads to low conversion efficiencies. It has been shown by Ito et al (8) that it is possible in some cases to maximise the overlap by using a multilayer guiding region.

The waveguiding structures used for SHG have been planar with the following structures: 1. a linear guide on a non-linear substrate (9-12); 2. a non-linear guide on a linear substrate (13-16); 3. a non-linear guide on a non-linear substrate (17-20). Recently, stripe waveguides have been fabricated by Ti indiffusion into a $LiNbO_3$ substrate for SHG (21-22). It has not been possible, up to now, to realise an efficient non-linear optical device using these structures due to a number of factors in addition to the constraints placed by the overlap integral. The most stringent requirement that has been difficult to achieve is the need to maintain uniform guide dimensions to within a few percent over the waveguide length in order to maintain phase matching. The other problems that have been serious are the scattering losses at the guide-substrate interface and the optical damage as a result of high optical intensities. It should be possible with careful design of waveguide structures and improved fabrication techniques to obtain greater SHG efficiencies than those obtained with bulk media.

Organic crystal cored fibres

As an alternative to planar waveguiding structures we report here the fabrication of crystal cored fibres in which it is possible to maintain uniform guide dimensions over long lengths. These fibres with organic crystal core material having large second order non-linearity could be used for miniaturization of visible laser sources and realization of parametric amplifiers for optical communications.

Organic crystal cored fibres have been fabricated by growth in glass capillaries of a single crystal from melt using a modified vertical Bridgeman technique. This method of crystal growth is only applicable to materials which are chemically stable on melting and have a melting point less than the glass transformation temperature. In addition, for optical guidance in these fibres it is necessary to select a cladding glass from commercially available glasses (23) having a refractive index less than the core.

Initial studies on the growth of organic crystals in glass capillaries were carried out by a number of researchers (24-27) to gain understanding of crystal growth mechanism in small bore capillaries and to identify useful materials for fabrication of fibre optic components. The growth of m-nitroaniline, m-dinitrobenzene, 2-bromo-4-nitroaniline, m-dihydroxybenzene, acetamide, benzil and formyl-nitrophenylhydrazine, has been carried out in 15μm to 80μm bore diameter glass capillaries. The best results for void free single crystal growth were obtained for m-dinitrobenzene, acetamide and benzil. With these materials, void free single crystal cored fibres of lengths upto 5cms have been obtained. Recently, similar lengths of void free m-dinitrobeneze and benzil crystals have been grown in glass capillaries with bore diameters in the 2μm to 10μm range for single mode operation (7,28).

For guided wave non-linear interactions in these fibres it is also necessary that the material be non-centrosymmetric, optically transparent in the wavelength region of interest and should have a non-zero SH tensor coefficient which is a multiplicative factor with one of the transverse electric field components. The latter requirement is necessary as the fields in a weakly guiding fibre are very nearly transverse (29). In the earlier studies on growth of m-dinitrobenzene and m-nitroaniline it was found that these crystals grow in glass capillaries with their crystal c-axis along the fibre axis. As these materials have their largest SH tensor coefficients along their c-axis and because the fields in the fibre are transverse they cannot be used for three wave mixing in this configuration.

For, SHG it is desirable to fabricate fibres which allow single mode operation at the fundamental wavelength so that all the launched power is available for conversion. The frequencies for which single mode operation is possible in a step-index uniaxial crystal cored fibre with crystal axis along the fibre axis are given by (30)

$$\omega \leq \frac{2.405c}{a(N.A.)}\, n_1/n_z \qquad : n_z > n_1$$

$$\omega \leq \frac{2.405c}{a(N.A.)} \qquad : n_z \leq n_1$$

where

a is the core radius.
c is the velocity of light in vacuum.
N.A. is the numerical aperature and is given by $(n_1^2 - n_2^2)^{1/2}$.
n_1 is the crystal refractive index transverse to the fibre axis.
n_z is the crystal refractive index along the fibre axis.
n_2 is the cladding glass refractive index.

Fabrication of benzil crystal cored fibres

In this initial exploratory study, benzil, a positive uniaxial crystal of trigonal class and point group 32 was used to demonstrate phase matched second harmonic generation in crystal cored fibres. Benzil's melting point is 95°C and good quality crystals can be grown with ease from its melt. The benzil's transmission spectrum, shown in Figure 1, was determined for a 5.49mm thick crystal grown from melt using a Perkin-Elmer spectrophotometer. This shows that benzil cannot be used for efficient frequency doubling of 0.9μm wavelength radiation. Benzil has two non-zero SH tensor coefficients, namely d_{11} and d_{14} of which d_{11} is of the larger magnitude (31). The magnitude of d_{11} coefficient and the SH coherence length have been measured by Jerphagnon (32) to be $(11.5 \pm 1.5) \times d_{11}$ of α-quartz and 7.1 ± 0.7μm respectively. Also, its refractive indicies have been measured in the 0.421μm to 0.656μm range (33) and Chandrashekar (34) has derived Sellmeier-Drude formuale to fit refractive index dispersion in this range. It has been experimentally verified by Jerphagnon that these equations can be used to give correct refractive index values upto 1.06μm wavelength and we have used them to design benzil cored fibres for SHG of 1.06μm wavelength radiation.

To fabricate single mode benzil crystal cored fibres for operation at 1.06μm wavelength with 6μm core diameter, Schott glass SKN18 was selected as the cladding glass. The capillaries, having bore diameters in the 2μm to 10μm range, were drawn from a suitably machined preform using conventional fibre pulling techniques. It was found that the bore diameter tended to be uniform over lengths of upto 10cm. The glass capillaries were filled with the crystal melt by capillary action in the 'hot' region of the furnace prior to the crystal growth. The 'hot' region of the furnace was maintained at a temperature of 5°C above benzil's melting point. The filling process takes only a few minutes for 5cm to 10cm long capillaries. The furnace used for crystal growth is essentially a resistance furnace with a water flow arrangement from a constant water head, shown in Figure 2. The furnace was well insulated in order to have a stable temperature profile. The crystal growth rate could be varied from 1mm/hr to 50mm/hr by means of a pulley having different radii connected to the shaft of a motor. The melt was

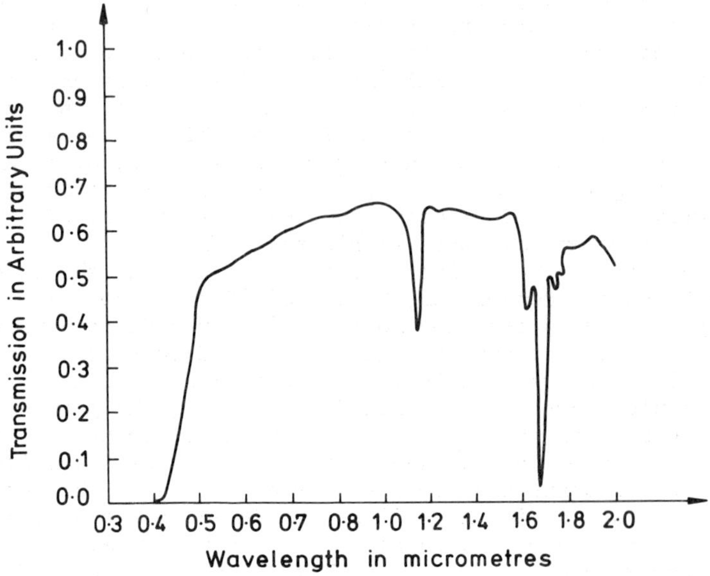

Figure 1. Transmission spectrum of bulk benzil crystal.

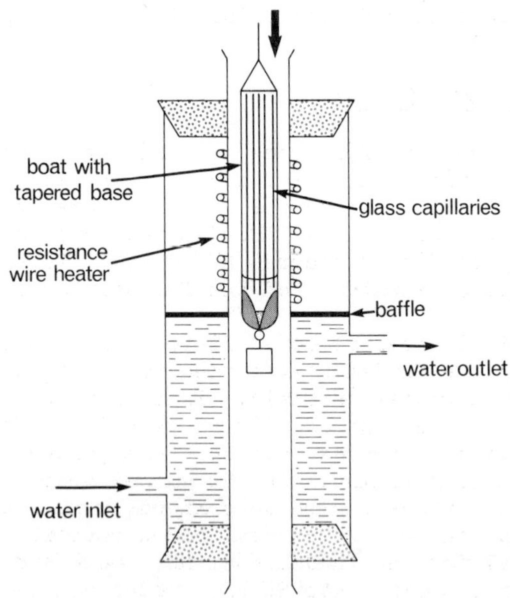

Figure 2. Crystal growth furnace.

then progressively crystallized by slowly moving the capillaries through a sharp temperature gradient.

Voids can occur on crystallization if there are bubbles present in the melt during filling. In addition, as crystallization occurs there is decrease in volume and if it is not filled by the flow of melt to the crystal face a void is generated. A pulling speed of 18mm/hr and a temperature gradient of 5°C/mm were found to give optimum conditions for void free single crystal growth. Void free monomode benzil crystal cored fibres of lengths upto 5cm have been fabricated. The crystal length was limited by the furnace design and, in principle, it should be possible to obtain longer lengths of void free fibre. A photograph of a void free monomode benzil cored fibre between the cross polarisers is shown in Figure 3. The direction of benzil crystal axis in glass capillaries was determined using a polarising microscope and was found to be along the growth direction which, in this case, was along the fibre axis. At times deviations of upto 5° were observed and these were due to the presence of transverse temperature gradients in the furnace.

The attenuation in these fibres was measured using a 0.633μm He-Ne laser and was estimated to be less than 2dB/cm. This value is only approximately 1dB greater than the attenuation in bulk crystal and hence implies good crystal quality. It was difficult to measure a more accurate value for propagation loss as it was not possible to determine the launch efficiency, strip cladding modes and obtain good end quality due to short lengths of the fibres fabricated.

SHG in crystal cored Fibres

The phase-matching for three wave mixing process can be achieved in an optical fibre for modes which satisfy the condition $\beta_1 = \beta_2 \pm \beta_3$, where β_1, β_2 and β_3 are the propagation constants for the three modes. For optical second harmonic generation this condition becomes $\beta^{2\omega} = 2\beta^{\omega}$, where β^{ω} and $\beta^{2\omega}$ are the propagation constants of the fundamental and the second harmonic.

Optical SHG in crystal cored fibres can take place by either coupling the fundamental guided mode, HE_{11}, into a SH guided mode or SH radiation field. These forms of phase-matching can be shown on a ω-β diagram, Figures 4 and 5. Here the refractive index differences have been exaggerated to show the principle of phase matching. The SH guided mode into which coupling can take place can be determined by the solution of the eigenvalue equation for the crystal cored optical fibre. The overlap integral for the fundamental HE_{11} mode and the SH modes is non-zero for coupling into SH HE_{1n} modes and is maximum for coupling into the SH HE_{11}. However, the waveguide dispersion characteristic for the HE_{11} mode does not allow phase matching between the fundamental and the SH HE_{11} modes and as a result the

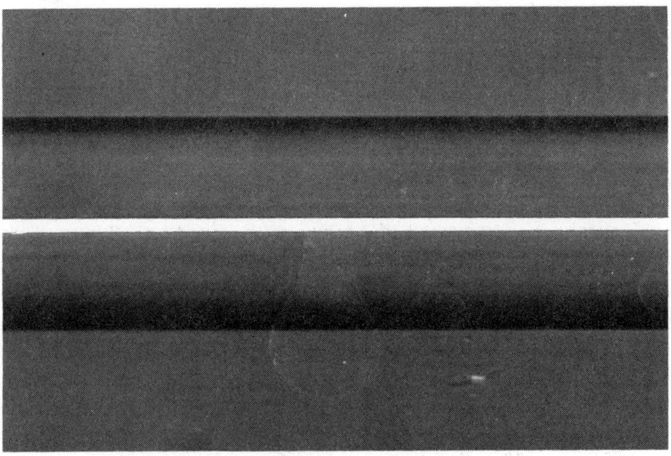

Figure 3. Monomode benzil crystal cored fiber viewed between cross polarizers.

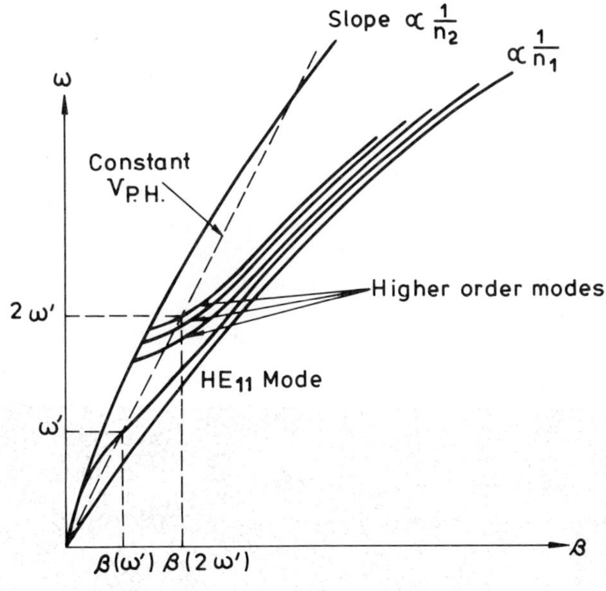

Figure 4. Diagram (ω–β) showing phase-matching scheme for guided wave optical SHG.

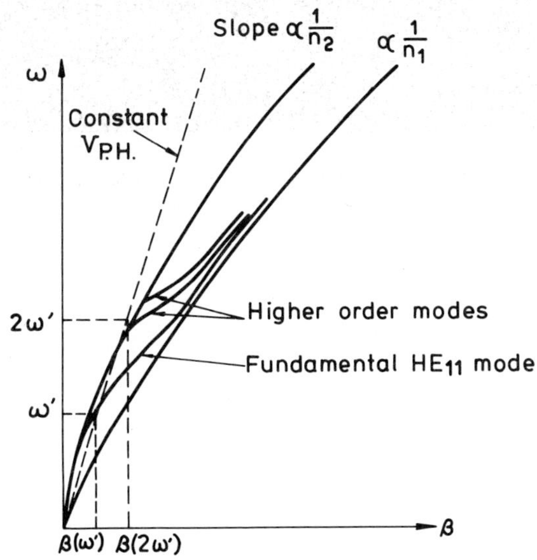

Figure 5. Diagram (ω–β) showing phase-matching scheme for optical SHG in the radiation field.

fundamental mode has to be phase matched to a higher order mode. Also, for guided wave non-linear interactions to take place, the phase matching requirement is not necessarily satisfied for a given waveguide dispersion. In some cases it should be possible to use the temperature dependent change in refractive index to change the waveguide dispersion and permit phase-matching.

The coupling to SH radiation field is simpler to implement and in this case it is only necessary to arrange that the propagation constant of SH, $\beta^{2\omega}$, is such that it lies in the continuum of the radiation field ie $\beta^{2\omega} < k^{2\omega} n_2^{2\omega}$, where $k^{2\omega}$ and $n_2^{2\omega}$ are the free space propagation constant and the cladding refractive index at 2ω. To implement this form of phase matching, the chromatic dispersion, δn, required in the cladding glass is given by

$$\delta n \geq [\beta^{\omega}/k^{\omega} - n_2^{\omega}]$$

where, $n_2^{2\omega} = n_2^{\omega} + \delta n$

The SKN18 glass for benzil crystal cored fibre has chromatic dispersion of 0.018 in the 0.532µm to 1.064µm wavelength range and this is sufficient for the SH to couple into the radiation field. The SH radiation in this case exits from the fibre core at an angle, α, given by

$$\mathrm{Cos}\alpha = \beta^{2\omega}/k^{2\omega} n_2^{2\omega}$$

where, $\beta^{2\omega} = 2\beta^{\omega}$, and β^{ω} is the propagation constant of the fundamental HE_{11} mode.

The angle 'α' in practice is of the order of few degrees and consequently the SH wave will be guided in the cladding glass. The theoretical analysis for this form of coupling shows that the SH conversion efficiency is proportional to the interaction length rather than the square of the interaction length as for the guided wave SHG. Also, in this case the overlap integral is small resulting in lower conversion efficiencies than the guided wave interaction.

Experimental results

In order to demonstrate optical SHG in benzil crystal cored fibres by coupling to SH radiation field, light from a Q-switched Nd:YAG laser operating at 1.064µm wavelength was launched into a 3cm long fibre using a x10 microscope objective. The fibre core diameter was approximately 3.75µm and the numerical aperture was 0.2. The SH so generated was seen to be guided in the cladding and could be seen on a screen to correspond to the far field pattern of the cladding, Figure 6. The SH conversion efficiency of 2×10^{-3} % has been estimated for this fibre.

To determine optical damage in bulk benzil crystals a Q-switched Nd:YAG laser with 1KW peak power, pulse width of 0.1μs and pulse repetition rate of 500Hz was used. The laser power was attenuated using a set of neutral density filters and focussed onto a bulk benzil crystal using a x10 microscope objective. No optical damage was observed with optical intensities of upto $100MW/cm^2$. Also, no optical damage was observed in benzil cored fibres with similar optical intensities.

Discussion

The SHG efficiency obtained in the above case is rather low. This is due to the small value of benzil's SH non-linearity and since the SH wave is guided in the cladding it may destructively interfere with the SH generated at later instants. This interference process is quite complex as the SH wave propagates in the cladding as a number of different modes, the cladding being a multimode dielectric tube waveguide, and the modes having similar field configuration and polarisation will interfere. This problem arises because of the long coherence length of the Nd:YAG laser and will not arise for instance with high power semiconductor lasers having short coherence lengths. Hence, coupling of SH to radiation field could be usefully employed for SHG in the visible using high power semiconductor lasers.

For a practical device, guided wave SHG and parametric amplification are preferable because of the higher conversion efficiency and the ease of use. It is possible to maximise the overlap integral by phase matching the fundamental and the SH HE_{11} modes by using either a biaxial crystal material or growing the crystal material in glass capillaries having an elliptical bore. In the case of a biaxial core material it may be possible to arrange that with the fundamental wave launched in the fibre as ${}_{y}HE_{11}$ mode, the phase matching condition is satisfied for coupling to the SH ${}_{x}HE_{11}$ mode. This form of phase matching is shown in Figure 7. In a similar manner it may be possible to exploit shape dependent birefringence of an elliptical cored fibre by having the fundamental and the SH modes as ${}_{o}HE_{11}$ and ${}_{e}HE_{11}$ respectively. Also, in order to use materials like m-nitroaniline, having large SH tensor coefficients along the crystal axis, the possibility of changing the crystal growth direction in glass capillaries is being investigated.

Conclusions

In conclusion a method of fabrication of guiding structure which is simple and suited to organic materials has been presented and optical SHG in benzil crystal cored fibres demonstrated. Also, phase matching schemes have been discussed for which the field overlap integral is maximum.

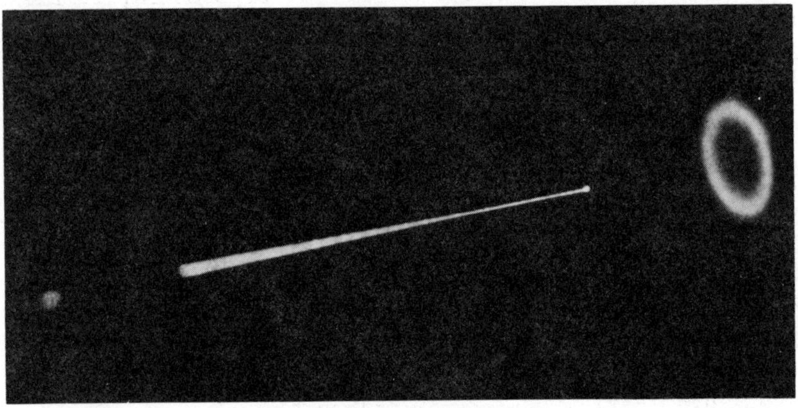

Figure 6. Optical SHG in benzil crystal cored fiber by coupling ω - β *to the radiation field.*

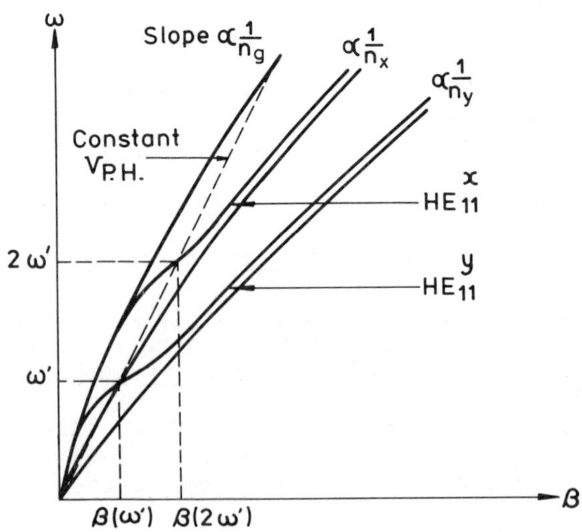

Figure 7. Diagram (ω–β) showing phase-matching scheme for guided wave SHG in a biaxial crystal cored fiber with both the fundamental and the SH as HE_{11} modes.

Acknowledgments

The author would like to thank J R Cozens of Imperial College, London, and R B Dyott of Andrew Corporation, Illinois, for useful discussions during the course of this work. Thanks are also due to R Kashyap of BTRL for supporting discussions and the Director of BTRL for permission to publish this paper.

Literature Cited

1. Carenco, A.; Jerphagnon, J.; Perigaud, A.J.Chem.Phys. 1977, 66, 3806-3813.
2. Oudar, J. L; Hierle, R.J.Appl.Phys. 1977, 48, 2699-2704.
3. Jain, K.; Hewig, G. H.; Cheng, Y. Y.; Crowley, J. I.IEEE J. Quantum Electron. 1981, 17, 1593-1594.
4. Lipscomb, G. F.; Garito, A. F.; Narang, R. S. J. Chem. Phys. 1981,75, 1509-1516.
5. Armstrong, J. A.; Bloembergen, N.; Ducuing, J.; Pershan, P. S. Phys. Rev. 1962 127, 1918-1939
6. Boyd, G. D.; Patel, C. K. N. Appl. Phys. Letts. 1966, 18, 313-315.
7. Nayar, B. K. Digest 6th Topical Mtg. on Integrated and Guided Wave Optics 1982, ThA2.
8. Ito, H.; Inaba, H. Optics Letts. 1978, 2, 139-141.
9. Tien, P. K.; Ulrich, R.; Martin, R. J. Appl. Phys. Letts. 1970, 447-450.
10. Chen, B. U.; Tang, C. L.; Telle, J. M. Appl. Phys. Letts. 1974, 25, 495-498.
11. Suematsu, Y.; Sasaki, Y.; Shibata, K. Appl. Phys. Letts. 1973, 23, 137-138.
12. Burns, W. K.; Lee, A. B. Appl. Phys. Letts. 1974, 24, 222-224.
13. Anderson, D. B.; Boyd, J. T. Appl. Phys. Letts. 1971, 19 266-268.
14. Zemon, S.; Alfano, R. R.; Shaprio, S. L.; Conwell, E. Appl. Phys. Letts. 1972, 21, 327-32 .
15. Ito, H.; Uesugi, N.; Inaba, H. Appl. Phys. Letts. 1974, 25, 385-38 . 16.
16. Van derZiel, J.; Illegems, M.; Fay, P. W.; Mikulyak, R. M. Appl. Phys. Letts. 1976, 29, 775-777.
17. Hopkins, M. M.; Miller, A. Appl. Phys. Letts. 1974, 25, 47-50.
18. Van der Ziel, J. P.; Miller, R. C.; Logan, R. A.; Norland, W.A. Appl. Phys. Letts. 1974, 25, 238-240.
19. Uesugi, N.; Kimura, T. Appl. Phys. Letts. 1976, 29, 572-574.
20. Uesugi, N., Daikoku, K.; Kubota, K. Appl. Phys. Letts. 1979, 34, 60-62.
21. Sohler, W.; Suche, H. Appl. Phys. Letts. 1978, 33, 518-520.
22. Uesugi, N. Radio Sci. 1982, 17, 197-204.

23. "Catalogue of optical glasses", Jenaer Glaswerk, Schott and Gen., Mainz (W.Germany).
24. Stevenson, J. L.; Dyott, R. B. Electron. Letts 1974, 10, 449-450.
25. Stevenson, J. L. J. Crystal Growth 1977, 37, 116-128.
26. Babai, F. H.; Dyott, R. B.; White, E. A. D. J. Mats.Sci. 1977, 12, 869-872.
27. Babai, F. H.; White, E. A. D. J. Crystal Growth 1980, 49, 245-252.
28. Ballentyne, D. W. G.; Al-Shukri, S. M. J. Crystal Growth 1980, 48, 491-492.
29. Gloge, D. Applied Optics 1971, 10, 2252-2258.
30. Gott, J. R. J. Phys. B: Atom. Molec. Phys 1971, 4, 116-123.
31. Jerphagnon, J. IEEE J. Quantum Electron. 1971, 7, 42-43.
32. Bryant, W. M. D. J. Am. Chem. Soc. 1943, 65, 96.
33. Chandrasekhar, S. Proc. Indian Acad. Sci. Sect. A 1954, 39, 243-253.
34. Cozens, J. R. Electron. Letts. 1976, 12, 413-415.

RECEIVED May 16, 1983

8

Optical Nonlinearities and Photoinduced Solitons in Conjugated Polymer Crystals

CHRISTOS FLYTZANIS

Laboratoire Propre du Centre National de la Recherche Scientifique (CNRS), Ecole Polytechnique, Laboratoire d'Optique Quantique, 91128 Palaiseau, Cedex, France

The linear and nonlinear optical properties of the conjugated polymeric crystals are reviewed. It is shown that the dimensionality of the π-electron distribution and electron-phonon interaction drastically influence the order of magnitude and time response of these properties. The one-dimensional conjugated crystals show the strongest nonlinearities ; their response time is determined by the diffusion time of the intrinsic conjugation defects whose dynamics are described within the soliton picture.
It is also shown that the electron-phonon interaction is operative in the polymerization process of the one-dimensional conjugated polymeric crystals ; a simple dynamical model for the polymerisation in polydiacetylenes is presented that accounts for the existing observations.

The potential use and applications of very sophisticated nonlinear optical processes in optical telecommunications and integrated optics have put requirements on the nonlinear optical materials which cannot all be met with the currently used inorganic crystals, ionic or semiconductors. The use of organic materials and polymers in particular became more imperative than ever. The interplay of high nonlinearity, appropriate form, size and response time of the nonlinear optical material which is very important in these applications can be more ingenuously optimized with the latter ones. Furthermore with a better understanding of the topochemical reactions that govern the growth and deposition of these materials new far reaching applications may be conceived by combining chemical and optical operations.

0097-6156/83/0233-0167$06.00/0

Theoretical estimations and experimental investigations firmly established (1) that large electron delocalization is a perequisite for large values of the nonlinear optical coefficients and this can be met with the π-electrons in conjugated molecules and polymers where also charge asymmetry can be adequately introduced in order to obtain non-centrosymmetric structures. Since the electronic density distribution of these systems seems to be easily modified by their interaction with the molecular vibrations we anticipate that these materials may possess large piezoelectric, pyroelectric and photoacoustic coefficients.

Conjugated polymer crystals are expected to occupy a priveleged position in diverse nonlinear optical devices if their topochemical growth is adequately mastered and their intrinsic defects are properly characterized. Here we assess the main factors and mechanisms that determine the magnitude and dynamics of the nonlinear optical coefficients in crystalline conjugated systems and stress the key role played by the dimensionality of the electron delocalization and the electron-phonon interaction. It is shown that these factors have their strongest impact in one-dimensional conjugated systems but with conflicting tendancies ; one-dimensional delocalization enchances the nonlinear coefficients while electron-phonon interaction in such systems introduces intrinsic conjugation defects that may limit their time response.

The polydiacetylene crystals (1-4) most strikingly corroborate these conjectures. Along this line of thought is also shown that this electron-phonon interaction is intimately interwoven with the polymerisation process in these materials and plays a profound role there. We make the conjecture that this occurs through the motion of an unpaired electron in a non-bonding p-orbital dressed with a bending mode and guided by a classical intermolecular mode. Such a polaron type diffusion combined with the theory of non radiative transitions explains the essentials of the spectral characteristics of the materials as well as their polymerisation dynamics.

Main structural and optical properties of conjugated polymers

Conjugated polymers are in general characterized by highly anisotropic optical, dielectric, conducting and mechanical properties (2,5). This is because the valence electrons, responsible for these properties, respond more easily to perturbations along directions where the conjugation occurs than in others ; their delocalization in the other directions is hindered by saturated bonds which keep the conjugated systems with an isolator type behavior in directions across the conjugation direction but semiconducting behavior along this direction. In the present review we shall limit ourselves in the semiconducting polymers.

The characteristic valence four of the C-atom is the origin of the enormous variety of these systems. Purely steric and topological considerations indicate that carbon conjugated polymeric structures can be either one- or two- dimensional since the conjugation arises from the electrons in p-orbitals which do not participate in sp-hybridization to form the skeleton of the saturated bonds. Indeed conjugation cannot be present if the four electrons are involved in four different bonding directions like in the diamond structure. Besides carbon, nitrogen or sulfur can also be used as building blocks of conjugated polymers but such materials are not yet fully characterized and will not be explicitly discussed. The different expressions for the nonlinear coeffcient derived below are however so parametrized that they may be used for other than carbon polymers as well.

Two-dimensional conjugated systems are thus expected to be formed by benzene type hexagonal rings so arranged to form plans and the π-electrons are delocalized there with only slight anisotropy ; these plans are well separated from each other so that the optical properties of these systems are quite different accross and parallel to the plans. Topologically we may distinguish (Figure 1) two structures the A and B, the later being non centrosymmetric ; the limiting case of the A-structure is the graphite (C) profably the only stable configuration. The refractive indices and optical gaps of graphite parallel and perpendicular to the plans are (6) $n_{\parallel} = 2.73$, $E_{\parallel} \simeq 0.0-0.1$ eV and $n_{\perp} = 1.53$, $E_{\perp} \simeq 6$ eV respectively ; its density $\rho \simeq 1.7 g/cm^3$. There are no measurements of nonlinear coefficients yet.

One dimensional conjugated carbon polymers can occur in many configurations as depicted in Figure 2 where also we included some chains with nitrogen and sulfur for later reference. Also included there are inorganic one dimensional semiconductors, like SbSI and SbSBr for later comparison. Besides the depicted one-dimensional systems, others like TCNQ- and KCP-salts could be included here as well but rough measurements of their nonlinear coefficients gave deceptively small values which combined with their ill-characterisation make them poor candidates for nonlinear optical devices.

Polydiacetylenes (2) come closest to the model one-dimensional organic semiconductor and can be readily obtained in form of large, nearly defect-free single crystals so that a large number of experiments and measurements have been carried out on these materials. Their structure is shown in Figure 3 where also some typical sidegroups R are indicated. On table I we summarize some measured va-

Table I. Optical properties of polydiacetylene crystals

Compound (R)	Ligth polar	Optical gap $E_o (cm^{-1})$	Density (gr/cm^3)	Refactive index n	$10^{12} \chi^{(3)}$ (esu)	Optical Breakdown Threshold (picoseconds) pulses
PTS	$\parallel$	15800	1.46	1.86	160	$10 GW/cm^2$
	$\perp$			1.58	<0.1	
TCDU	$\parallel$	17900		1.80	40	
	$\perp$			1.65	<0.1	$3 GW/cm^2$

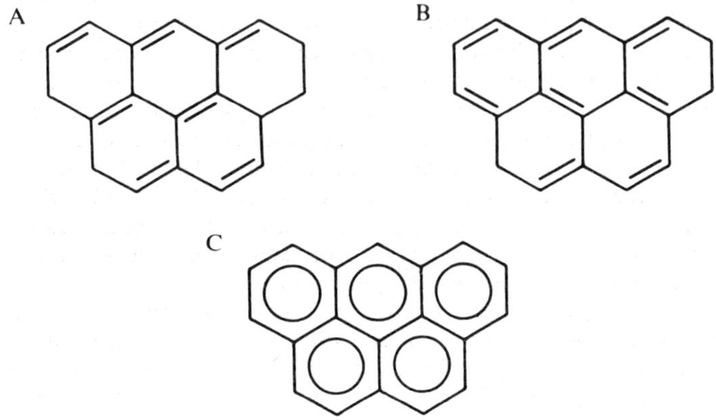

Figure 1. Two-dimensional structures. See *text for discussion.*

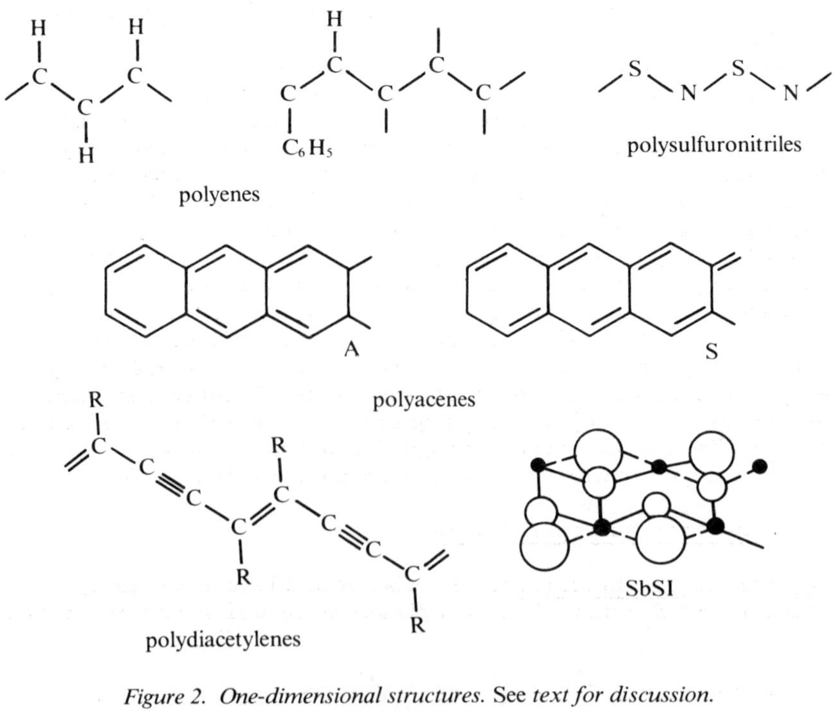

Figure 2. One-dimensional structures. See *text for discussion.*

lues (3) of coefficients relevant for optoelectronic devices. It should be stressed that these values may not be indicative for all polydiacetylenes as they strongly depend on the side groups which drastically affect their structural characteristics.

As a general rule these crystals exhibit a sharp absorption edge and a strong peak in reflectivity at roughly 2eV for light polarized parallel to the chains (7). Their spectrum along the same polarisation contains additional sidebands indicative of strong electron-phonon coupling ; on the other hand the spectrum perpendicular to the chain direction is very weak and featureless up to 4 eV. Optical and Raman spectroscopy (resonant and nonresonant) down to helium temperature revealed (7-8) dramatic changes in the spectrum of some polydiacetylenes (figure 4).

Besides their potential use as optical devices the interast in these materials relies on the fact that they are obtained through a unique topochemical polymerization which proceeds from the monomer crystal by a very specific motion of the monomer on its lattice site imposed by the packing of the molecules ; polymerization occurs by successive tilting of each monomer along the ladder without moving the center of gravity. It can be caused by X-or UV-irradiation, heating of shear stress. The thermal polymerisation of PTS most dramatically exhibits an autocatalytic effect (9) in the conversion-vs time curve which sets in at about 10% polymer conversion, the activation energy however is the same throughout the process (∿ 22 kcal/mole). A key role in the polymerisation process is played by a 5% mismatch between monomer and polymer repeat distances in the chain direction (a polymer is required to elongate by a 5% to be commesurate with the monomer lattice). There are ESR measurements indicating (10) formation of biradical and carbene-intermediates. These measurements essentially concern the two compounds PTS and TCDU (Figure 2). All polydiacetylenes do not exhibit such well defined properties and it is doubtfull whether their majority does. There is a gap to be bridged between the actual chemical achievement and the desirable ones for physical investigation and engineering applications. Furthermore their defects are not yet properly characterized. The situation is still worse for other types of crystalline conjugated polymers. Time resolved studies are needed (11) in the picosecond range for a better understanding of their optical properties and intrinsic photoexcited defects.

Nonlinear optical susceptibilities

Definitions.Expressions. The electric field E of an optical beam induces in a material a polarization ΔP which can be written (12) :

$$\Delta P = P - P_0 = \chi^{(1)}E + \chi^{(2)}EE + \chi^{(3)}EEE + \ldots \qquad (1)$$

for stationary fields on the transparency region of the material ; it is assumed that the electric field causes no irreversible chemi-

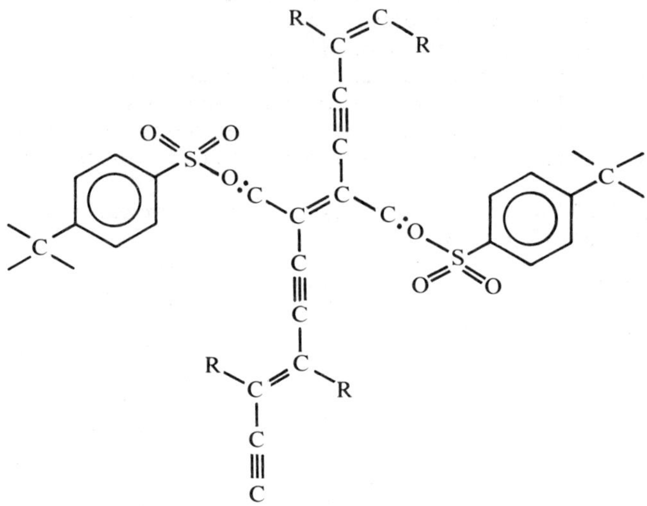

Figure 3. PTS-diacetylene crystal.

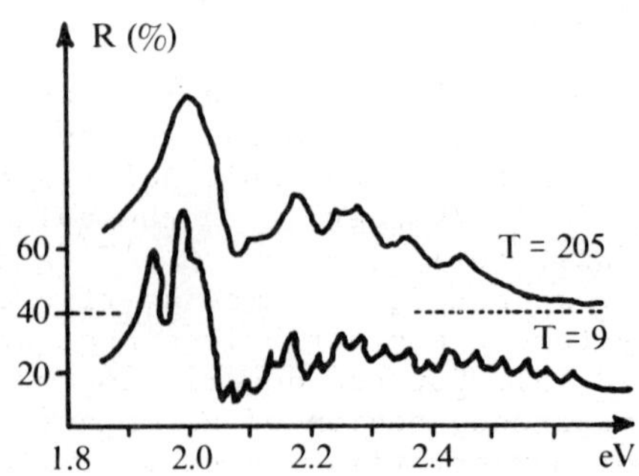

Figure 4. Reflectivity of the PTS-diacetylene crystal for two different temperatures.

cal changes in the material. The susceptibilities $\chi^{(n)}$, linear and nonlinear, are tensorial quantities that measure the polarizing effect of the optical fields on the outer valence electrons. Since the later are the most strongly affected by the chemical bonding of the material the order of magnitude of the susceptibilities varies substantially from material to material ; in particular $\chi^{(2n)}\equiv 0$ for any centrosymmetric material . It should be kept in mind that in setting expression (1) it is tacitly assumed that the matter-radiation interaction is local (dipolar).

In table II we have gathered some representative values of inorganic materials and in table I and III those of organic ones. To make the comparison easier we have also inserted the corresponding

Table II. Nonlinear coefficients of some inorganic semiconductors (reference 13)

Compound	$10^{6}\chi^{(2)}$(esu)	$10^{30}\beta$(esu)	$10^{12}\chi^{(3)}$(esu)	$10^{36}\gamma$(esu)
GaAs	0.9	22	12	120
InSb	3.3	110	>50	>1700
Si	0	0	6	60
Ge	0	0	100	2300
SbSI	1	-	-	-

Table III. Nonlinear coefficients of some organic compounds

Molecule	$10^{30}\beta$(esu)	$10^{36}\gamma$(esu)	ref.
Trans β-carotene	-	10^{3}	14
Trans retinal	-	10^{2}	14
CH_3-N⬡CH=CH⬡NH_2	42	13	15
NO_2⬡NH_2	21	5	16
CH_3-N⬡CH=CH⬡O	10^{3}	-	17

values of inorganic materials and in table I and III those of organic ones. To make the comparison easier we have also inserted the corresponding values of the macroscopic polarisabilities $\alpha^{(n)}$ formally defined by $\chi^{(n)}\equiv\alpha^{(n)}/v$ where v is the value of the repeat unit in the crystal (unit cell) ; we put $\beta\equiv\alpha^{(2)}$ and $\gamma\equiv\alpha^{(3)}$. Let us evaluate the order of magnitude of these coefficients.

In periodic systems the electronic states are distributed in Bloch type bands (18,19). For the description of the optical properties in the transparency region of an organic semiconductor to which conjugated polymer crystals can be assimilated we may limit ourselves to two bands, the highest valence (v) and the lowest conduction (c) bands. Using the Genkin-Mednis (20,12) approach, one can derive the expressions of the nonlinear susceptibilities as integrals of certain combinations of inter- and intraband terms over the whole Brillouin Zone (or eventually the extended Jones Zone (18,19)). Thus for the xx-components of the linear and third order susceptibilities of a centrosymmetric semiconducting crystal one has (4,21,22)

$$\chi^{(1)}_{xx} = \frac{4e^2}{\hbar V}\int_{B.Z.} \Omega_{vc}\, S_{cv}\, dk \qquad (2)$$

$$\chi^{(3)}_{xxxx} = \frac{8e^4}{\hbar^3 V}\int_{B.Z.}\left[\frac{1}{\omega_{cv}}\frac{\partial S_{cv}}{\partial k}\frac{\partial S_{vc}}{\partial k} - \Omega_{vc}S_{cv}S_{vc}S_{cv}\right]dk \qquad (3)$$

where the band energy difference $\hbar\omega_{cv}=\varepsilon_c-\varepsilon_v$, Ω_{vc} is the transition dipole moment matrix element between the two bands, $S_{vc}=\Omega_{vc}/\omega_{vc}$ and $\varepsilon_{c,v}$ are the band energies. It is important to notice that the linear susceptibility $\chi^{(1)}$ only arises from interband transitions, while $\chi^{(3)}$ arises from the competition of interband and intraband terms and this is true for all nonlinear susceptibilities ; for very delocalized systems (namely strong overlap of wave functions) the quantities ω_{cv} and Ω_{vc} vary strongly with k over the Brillouin Zone and the intraband term becomes the dominant one (4).

From the physicochemical point of view, the prevailing one in the present review, the question is how these coefficients relate to the characteristics of the constituant molecules and how this information can be extracted from (2,3). The answer relies on the fact that the essential contributions to these integrals come from only few nonoverlapping critical regions in the joint density of states (18,19) ; these are points, lines and surfaces depending on the spatial extension of the conjugated electron distribution. They are defined by the condition

$$\vec{\nabla}_k \omega_{cv} = 0 \qquad (4)$$

This directly establishes a relation between the optical susceptibilities $\chi^{(n)}$ and the topology and dimensionality of the joint density of states ; more importantly it allows one to express $\chi^{(n)}$ in terms of the values that Ω_{cv} and ω_{cv} take at these critical regions; the latter being expressed in terms of the characteristics of the electron distribution (heteroatomicity, bond alternation and superalternation, chain pairing, electron correlation) this functional dependence allows one to globally assess the impact of these features on the nonlinear susceptibilities.

One-dimensional crystals. In one-dimensional systems (1D) the problem is amenable to an analytical solution and allows one to gain some insight to this approach. We illustrate this point with the simple bond alternated infinite carbon chain (polyene) and adopt the Hückel approximation. Such a chain is then characterized (23) by two resonance energies β_1 and β_2 that alternate along the chain. We assume complete separation between σ- and π- electrons and we concentrate our attention only on the π-electron contribution. The electron states in such a chain can be obtained analytically and one has (see 21, 22) $h\omega_{cv}=2\beta_2\zeta_0$, $e\Omega_{cv}=iea(1-\nu^2)/4\zeta_0^2$ where $\zeta_0=(1+\nu^2+2\nu\cos ka)^{1/2}$ and $\nu=\beta_1/\beta_2$. Inserting this expression of ω_{cv} in (4) we see that the joint density of states becomes infinite at the edge of the B.Z., ka=π (critical point), where also the matrix element $\Omega_{cv}(ka)$ attains its maximum value (see Figure 5)

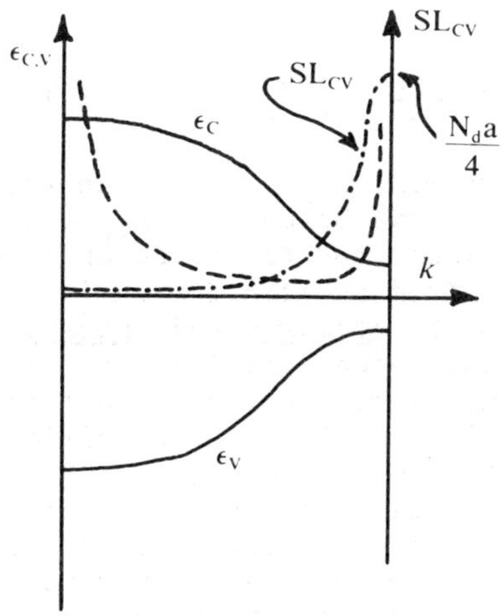

Figure 5. Energy bands (—), dipole transition moment (-•-), and density of states (---) for bond alternated chain.

$|\Omega_{cv}(ka=\pi)|=a|\beta_1+\beta_2|/|\beta_2-\beta_1|\equiv L_d$ (optical delocalization length) and the energy difference $\hbar\omega_{cv}$ becomes smallest $\hbar\omega_{cv}(ka=\pi)=2|\beta_2-\beta_1|=E_o$ (optical gap). The coincidence of these three features at a single point is an essential characteristic of the 1D systems and is the origin of the enhancement of the π-electron contribution over the σ-electrons in $\chi^{(n)}$. Indeed one finds (4,22) that for strong delocalisation, $N_d\equiv 4L_d/a>1$, the intraband term in (4) is the dominant one and

$$\chi_\pi^{(1)} = \frac{2}{3\pi}\chi_\sigma^{(1)}N_d^2 \qquad (5)$$

$$\chi_\pi^{(3)} = \frac{16}{45\pi}\chi_\sigma^{(3)}N_d^6 \qquad (6)$$

and in general

$$\chi_\pi^{(2n-1)} = \chi_\sigma^{(2n-1)}N_d^{4n-2} \qquad (7)$$

where $N_d=|\beta_1+\beta_2|/|\beta_1-\beta_2|$ and $\chi_\sigma^{(n)}$ is the expression of the susceptibility for a chain of saturated bonds (σ-electrons). It has been shown (4,22) that the above functional dependence of the odd order susceptibilities on the delocalization parameter N_d is very general and characteristic of all 1D systems (with or without center of inversion, chain pairing or electron correlation). Since N_d can attain high values for conjugated polymers the much higher dependence of the nonlinear susceptibilities on N_d than the quadratic one found for $\chi_\pi^{(1)}$ justifies the neglect of the side group contribution to $\chi_{xxxx}^{(3)}$ while this is not necessarily the case for $\chi_{xx}^{(1)}$; there the contributions of the σ- and π-electrons to $\chi_{xx}^{(1)}$ are found to be of the same order of magnitude. Thus the nonlinear susceptibilities can be effectively used to measure the conjugated electron delocalization. Local field corrections can also be taken into account (21,22).

Expressions (5-7) can also be cast in the form

$$\chi^{(2n-1)} \sim \chi_\sigma^{(2n-1)}\left(\frac{E_f}{E_g}\right)^{4n-2} \qquad (8)$$

which contains directly accessible experimental values (E_F is the Fermi level and E_g is the optical energy gap) and also allows direct comparison with inorganic semiconductors SbSI and SbSBr which can have high values of nonlinearities as well since $E_F/E_g>>1$ there; unfortunately the crystal growth for these materials is at presently an insurmontable problem (24).

The situation is drastically different with the second order susceptibility which is non-vanishing only in materials without inversion symmetry. Its expression is (22)

$$\chi^{(2)}_{xxx} = \frac{i6e^3}{\hbar^2 V}\int_{B.Z.}\left\{S_{vc}S_{cv}(\Omega_{vv}-\Omega_{cc}) - \frac{1}{2}\left[S_{vc}\frac{\partial S_{cv}}{\partial k} - \frac{\partial S_{vc}}{\partial k}S_{cv}\right]\right\}dk \qquad (9)$$

Here the interband and intraband terms can each have either sign so that they may add or substract. It is intructive so study again the bond alternated carbon chain in the Hückel approximation ; in order to be asymmetric hetero-atomicity must be present as well which is taken into account by letting the coulomb integral to take alternatively the values α_1 and α_2 along the chain. The expressions of Ω_{vc} and ω_{cv} are given in (21,22) . Careful analysis of (9) reveals that contributions to the integrand only come from regions where Ω_{cv} is complex ; it vanishes whenever Ω_{vc} becomes real or pure imaginary and this is precisely what happens at the edge of the Brillouin zone of an 1D system. Thus in contrast to $\chi^{(3)}$, $\chi^{(2)}$ does not take full advantage of the infinite density of states there and the highly delocalized character of these states ; or otherwise stated, charge assymetry, (a perequisite for non-vanishing $\chi^{(2)}$), being a local feature shows its full impact at most over a unit cell and larger delocalization is irrelevant. Expression (9) with the commonly used values of the Coulomb and resonance integrals, (α_1,α_2) and (β_1,β_2) respectively, shows that $\chi^{(2)}$ cannot exceed (22) the value of 10^{-7} esu which is an order of magnitude below that of GaAs and two orders of magnitude below the one expected for a merocyanine crystal (17); this seems to be confirmed from recent measurements in asymetric polydiacetylenes (25). Therefore any improvement of the values of $\chi^{(2)}$ ought to be sought in the optimisation of the polarizability values of small molecules (like the merocyanine) and their packing in the crystal. In particular side-groups with large second order polarizability values appropriately attached along a polydiacetylene chain may provide crystals with large values of $\chi^{(2)}$ along directions other than the chain direction ; the crystal growth of such materials seems however difficult if not hopeless (25). The 1D inorganic semiconductors SbSI and SbSBr (see Figure 2 and Table II) seem also to be favorable materials with large values of $\chi^{(2)}$ (24) ; unfortunately here too the crystal growth seems at presently an insurmontable problem.

Quite intriguing is also the case of centrosymmetric homoatomic bond alternated chains which get coupled into pairs in an asymmetric configuration (A). For such a system $\chi^{(2)}$ can attain (22) values larger than those of the single heteroatomic bond alternated chain.

Two- and three-dimensional crystals. In two-dimensional (2D) systems the critical regions are of two kinds : points at energy E_0 at the edge of the B.Z. reminiscent of 1D patterns in the electron distribution with a contribution in $\chi^{(1)}$ and $\chi^{(3)}$ as given by (7) and lines at energy E_1 intrinsic to the 2D system. The effect of the latter in the nonlinear susceptibilities is to reduce the impact of the point E_0 by borrowing and diffusing its oscillator strength and density of states and repelling E_0 to higher energies ($E_1 < E_0$). The analytical calculation of (3) even in the Hückel approximation is quite involved although the expressions of Ω_{cv} and ω_{cv} can be readily derived ; theoretical calculations in graphite confirm the above qualitative conclusion giving values of $\chi^{(3)}$ lower by one or two order of magnitude than for 1D systems. In general planar systems are rather unstable and their use as nonlinear optical devices is questionable.

Three-dimensional (3D) conjugated systems do not exist but one can conceive 3D organic semiconductors comparable to the inorganic ones (Ge, GaAs). There, besides contributions from points E_0 and lines E_1, $\chi^{(n)}$ will also contain contributions from critical surfaces at energy E_2. Since $E_2 < E_1 < E_0$ the impact of E_0 is further reduced. This is strinkingly evidenced in inorganic semi-conductors where estimations of the different contributions have been made using (26) simple models for the joint density of states. It is found that the main contribution to $\chi^{(3)}$ comes from the 2D and 1D critical regions of the B.Z. while the contribution from the critical point E_0 (zero dimensional) the so-called Penn-gap is substantially reduced as one proceeds to low gap semiconductors. This is because the intraband terms are more important in the lower energy critical regions than in the higher ones (Ω_{cv} and ω_{cv} vary more rapidly over the B.Z.. On the other hand a semiconductor like Ge, with one of the largest values of $\chi^{(3)}$, as will be discussed below, barely equals that of a polydiacetylene. This is particulary striking when contrasted with the fact that the density of valence electrons in Ge is by an order of magnitude higher than that of the π-electron in a polydiacetylene. Furthermore the polydiacetylenes have a higher transparency, are optically anisotropic and possess a lower index of refraction than Ge. It is unfortunate that there is no measurement of $\chi^{(3)}$ in inorganic 1D semiconductors (24) like SbSI and SbSBr since they would allow a better comparison with the organic semiconductors.

Polydiacetylene crystals. The enhancement of $\chi^{(3)}$ because of one-dimensional electron delocalization is strikingly corroborated in the polydiacetylene crystals. Their structure is that of a super alternated chain with four atoms per unit cell and the Hückel approximation yields four bands for the π-electrons, two valence and two conduction bands. When depicted in the extended Jones zone, each pair can be viewed as arising by a discontinuity at the middle of the Brillouin zone of the polyene chain. The dominant contribution to $\chi^{(2n-1)}$ comes from the critical point at the edge of the extended Jones zone (initially at the center of the reduced B.Z.). The complete expressions are derived in (4,22) and calculated for different polydiacetylenes. We reproduce the values of $\chi^{(3)}$ for TCDU and PTS in table IV. The calculated values are in good agreement with the measured ones (3). They are of the same order of magnitude as the values of $\chi^{(3)}_{\xi\xi\xi\xi}$ in Ge and GaAs which are 10^{-10} and 10^{-11} esu,

Table IV. Calculated values of $\chi^{(3)}$ in time polydiacetylenes (Ref. 22)

Compound (R)	Light polarization	Refract Index	$10^{12}\chi^{(3)}$ (esu)
PTS	‖	1.92	70
	⊥	1.72	<.1
TCDU	‖	1.76	25
	⊥	1.65	<.1

respectively (Table II), ξ being one of the (111)-crystal directions. These high values of $\chi^{(3)}_{xxxx}$ are due to the high delocalization of the π-electrons, a consequence of the large conjugation that is established upon polymerization. This is clearly seen when one compares the value of $\chi^{(3)}$ in the polymer with that of the monomer ; in the latter case $\chi^{(3)}$ has values less than 10^{-13} esu. The same is true for the values of $\chi^{(3)}$ across the chains in the polymer crystal ; these are less than 10^{-13} esu. These low values can be well accounted for by using additivity arguments for the hyperpolarizability. This dramatic increase of $\chi^{(3)}_{xxxx}$ over the other components because of the π-electron delocalization is particularly striking when contrasted with the facts that (a) the density of the localized electrons, which are the only ones contributing to the other components, is much higher than that of the π-electrons (only 1 out of 40 valence electrons per unit cell is a π-electron) and (b) their contribution to $\chi^{(3)}$, across the chain direction, is further enhanced by local field corrections (20,22). Because of the much milder dependence of $\chi^{(1)}_{xx}$ on N_d, the linear susceptibility does not show such marked anisotropy. The above values of $\chi^{(3)}$ where measured (3) at room temperature. At lower temperatures, below 160°K, the optical properties of some of these polymers, like the PTS-diacetylene, undergo some dramatic changes (absorption peak splitting, etc...) and the conjecture was made (27) that the chains may get coupled in pairs in an S-configuration at low temperature. The mechanism of the pairing is probably a rearrangement of the side groups allowing the π-electrons of one chain of the pair to jump over to its partner chain. If this is the case, the considerations of ref (22) indicate that $\chi^{(3)}$ may increase by an amount outside the experimental error.

There is no experimental determination of the sign of $\chi^{(3)}$ yet; this can only be obtained by interference techniques (CARS or similar (12)). In the tight-binding (or Hückel) description it is predicted to be plus (+) in the large delocalization limit (expression 6) and minus (-) in the small delocalization limit (flat bands). On the other hand in the exciton description occasionnally adopted by some authors to interpret the main absorption peak in the polydiacetylenes one finds $\chi^{(3)}$ negative and its values two orders of magnitude lower than expression 6 ; since electron correlation (28) is essential in the exciton model, the calculation of even the simplest optical properties becomes prohibitively complicated and the physical insight is obscured.

If any of the frequencies (ω, 2ω or 3ω in third harmonic generation or similar frequency combinations in three wave mixing experiments) approach the region of intense absorption situated around 16000 cm^{-1} appreciable changes in the values of $\chi^{(3)}$ are expected (22) because of the dispersion and have been observed (3, 29). In particular the intrinsic two-photon absorption without taking into account electron-lattice coupling was predicted (22) to be important and some experimental evidence (30,31) seemingly corroborate this prediction but the precise origin of this absorption is still unclear. Actually the electron-phonon coupling in 1D conjugated chains leads to drastic modifications in the electron states at the edge of the B.Z.. Thus Fano-type interference (32) between the continuum of band states and vibronic ones significantly changes the appearance of the absorption spectrum above the main peak. More importantly, the main absorption peak is strongly coupled with the distorsions of the ⟋C⟍ unit giving rise to a polaron type state. This introduces an inertia and limits the response speed of the optical nonlinearities as will be shown in the next section.

Time response of the optical nonlinearities. Solitons.

The most striking implication of the electron lattice coupling in 1D chains is the appearance of the semiconducting state : the equal bond 1D lattice (metallic state) is unstable (33) with respect to a lattice distorsion and this so called static Peierls instability is the origin of the opening of the intrinsic band gap at the edge of the B.Z. with an infinite density of states there and the presence of band alternation.

By extension one may say that the power laws (5-7) which determine the magnitude of the linear and nonlinear optical coefficients are consequences of this strong electron-lattice coupling. We now make the conjecture that the time response of these coefficients is severely affected by the dynamics of the electron-lattice coupling in conjugated chains when two or more resonant chemical structures can coexist ; this is the case for many of the organic chains of Figure 2.

Because of this degeneracy and coexistence of these resonant structures one expects intrinsic conjugation defects to exist and be formed along the chain, the so-called Pople-Wamsley defects (34) which in the case of polyacenes and polydiacetylenes takes the form

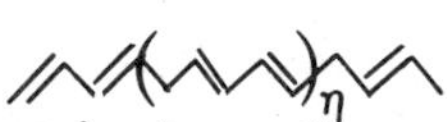

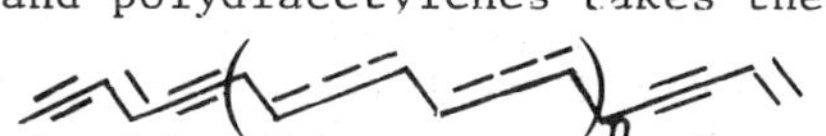

respectively ; they are characterized by the appearance of new states below the band gap E_0 that borrow their oscillator strength from there. It is the formation and motion of these defects during the light pulse that determine the temporal behaviour of the optical nonlinearities.

Since 1D conjugated chains are interrupted by such intrinsic defects (the Pople-Wamsley defects) or eventually impurities the

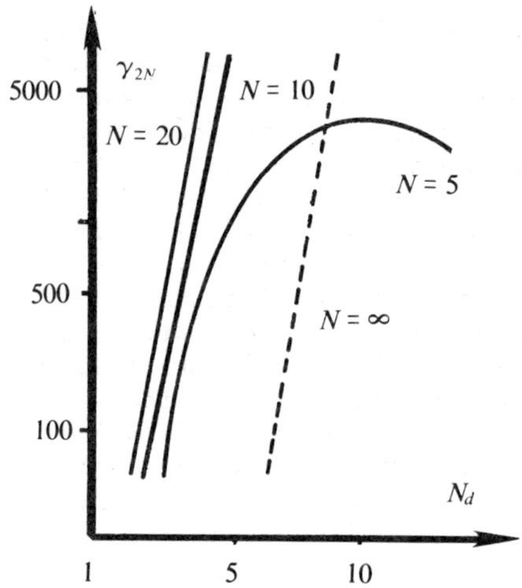

Figure 6. γ_{2N} vs. N_N for finite chains of different lengths. The dotted line corresponds to the slope of $\chi^{(3)}$ for an infinite chain.

first question arises then to what extend the behaviour of $\chi^{(3)}$ obtained in (6) for an infinite chain persists down to finite chains of 2N sites ; we neglect momentarily any motion of these defects. The relevant optical coefficient is now the third order polarizability γ. It is shown in (4) and depicted in Figure 6 that as long $N_d<N$ one gets $\gamma \sim N_d^6$, i.e. the same functional dependence as for an infinite semiconducting chain but for $N<N_d$ and N large, one gets (1,22) $\gamma \sim N^5$ which is characteristic of a free electron behaviour. These considerations clearly show the crucial role played by the electron delocalization length L_d for the magnitude of the nonlinear optical coefficients; its dynamical behaviour, dictated by the soliton type motion (35,36) of these defects, also determines the response time of these defects.

These defects can be create by photoexcitation of frequency ω_1 near the main absorption peak but also below it. The strong electron-lattice coupling leads to a polaron type or soliton state (35,36) where the electron state is dressed with large amplitude bending mode extending over a few repeat units and propagating roughtly with the sound velocity. This introduces a time scale of roughtly 1-4 ps right after photoexcitation during which the appearance of the absorption peak and the polarisation state of the chain will be markedly different from that observed for pulses longer than 10-100 ps. This behaviour will strongly be felt by a second pulse of frequency ω_2 below the band gap and this is precisely the case in three wave mixing experiments described by $\chi^{(3)}(\omega_1, -\omega_1, \omega_2)$.

For pulses shorter than $\sim$ 1ps the sepctrum will be broader and less pronounced that for pulses longer than 1 ps. Otherwise stated, the electron states immediately after excitation find themselves in a "foreign lattice" and this introduces strains which take finite time to diffuse before the π-electrons accomodate themselves in a new lattice. These initial strains broaden the spectrum in the initial stage of excitation (up to 1 ps) ; for time longer than 1 ps the spectrum coalesces to the one observed with very long pulesor stationnary sources.

Thus the π-electron delocalization length L_d and the linear and nonlinear optical coefficients in 1D conjugated systems reach their full values computed in the previous section in times longer than 1-4 ps, the diffusion time of these initial strains. This finite response time of the optical coefficients in 1D conjugated systems should be contrasted with the essentially instantaneous one expected in 3D semiconductors ; in 1D conjugated systems the π-electrons cannot avoid the contraints along the chain and are forced to accomodate with the latter and eventually loose some of their characteristics. The dynamical behaviour of the optical linear and nonlinear coefficients along this approach is derived in 37.

The above conclusions introduce intrinsic limitations to the use of the 1D conjugated systems in nonlinear optical devices. Although these may benefit (38) from the high nonlinearities, their response speed will be limited by the motion of such defects. These may also be formed by other means than light and this will clearly have implications on photoelastic, pyroelectric and piezoelectric effects as well. We point out that materials like polydiacetylenes may show appreciable quadrupolar pyroelectric effect (39).

Selective phonon localization and polymerization of diacetylenes

The solid state polymerisation of diacetylenes (2) with U.V. radiation, heating or shear force is most indicative of the predominant influence of electron-lattice coupling. The details of the chemical changes that occur during the polymerisation process are crucial (2,40) but the overall description only needs part of this chemical information. The kinetics and thermodynamics of the polymerisation process using an elastic strain approach have been worked out in (41).

Here we outline a dynamical description (42) of the polymerisation of the polydiacetylenes. The approach relies much on the one used (43,44) in the theory of non radiative transitions in crystals and the soliton description of the defects in the 1D-organic semiconductors.

One of the basic assumptions of this theory is that the polymerisation rate can be computed from the transition rate from an initial electronic state E_i to a final one E_f of the crystal at a given polymerisation state. The energies of these states depend on the nuclear configuration and their changes around the equilibrium positions for the initial and final electronic states can be expressed (43) in terms of vibrational oscillators which at a given temperature are either classical $\hbar\omega_c<kT$ (predominantly intermolecular) or quantum $\hbar\omega_q<kT$ (predominantly intramolecular).

The overall energy being conserved the transition form the initial to the final state occurs at a configuration (the transition state) where the two states, E_i and E_f, have the same energy. The associated activation energy E_a and reaction path are determined by the classical modes the main contribution coming from a very specific intermolecular mode (guiding mode) and an almost intramolecular one, the C≡C-C bending mode at approximately $\omega_b<300\ cm^{-1}$ which is strongly coupled with the adjacent electrons of the triple bond ; at room temperature this mode is in the dividing line between classical and quantum mode ($h\omega_b \simeq kT$). It is easy to see that when this bending mode is driven beyond a certain amplitude a non bonding p-orbital is formed with one unpaired electron at chain-end which is very reactive; the biradicals and carbenes, the chemical intermediates (40), are precisely of this form. The guiding intermolecular mode $\hbar\omega_c$ then creates the favorable conditions for the reaction with the neighboring species to take place. Clearly the electronic states involved are different at different stages of the polymerisation process. In the initial stages, (formation of oligomers from monomers) the lifetime of the reactive species is short but beyond a certain polymerisation, (by progression and fusion of chains) a stabilisation of the reactive species occurs which now can be pictured as an unpaired electron in a non-bonding p-orbital dressed with a large amplitude bending mode and guided through a classical intermolecular mode. This situation strongly reminds that of a polaron (35). With the polymerisation progressing to infinite chains the butatriene structure 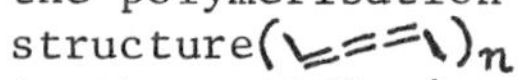 more favorable for the strained chain evolves to the acetylynic one  as the strains are released. This is also the origin of the intrinsic conjugation defects. When in particular, the polymer is photoexcited the butratiene structure is again favorable energetically and one has a large density of such defects at the initial stage of the photoexcitation for a time shorter than ∿1-4ps; thus, one may say that at this initial stage of photoexcitation the polymer rememorates the final polymerisation stages and one may even expect hysteresis effets.

The above dynamical description of the polymerisation strongly parallels that of nonradiative transitions and this is not accidental ; althouth the monomer crystal from which the polymeric one is issued, do fluoresce, the polymeric one does not, despite its strong absorption at 2 eV. This strongly indicates efficient nonradiative relaxation of the excitation and strong electron-phonon coupling.

The above picture points to the very interesting possibility of selectively inducing or enhancing the polymerisation process, at a temperature where this is unlikely, by resonantly driving with an intense laser beam in the infrared the vibrational modes ω_b and ω_c that are involved in the polymerisation. As a consequence of their anharmonicity (45) these modes, when driven near resonance by an electromagnetic field, beyond a certain critical value of the later, can reach amplitudes comparable to the critical ones required for the polymerisation to be initiated or proceed ; the anharmonicity in the presence of the intense laser beam acts as a defect and localizes the phonons creating thus a critical distorsion.

We wish to conclude with a simple kinetic model that is compatible with the above picture and is based on the theory of continuum Zip-reactions (46,47). Let us assume random initiation at sites distributed at random along a chain followed by zip in both directions with speed v. If we denote by ρ the linear density of such

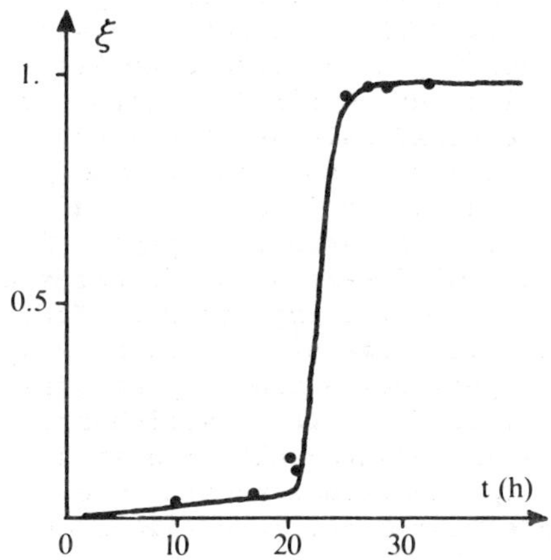

Figure 7. Calculated conversion rate (—) at T = 60 ° from expression 11 (see *text) with* λ *= 0.03 h⁻¹, t_c = 20 h 30 min, ξ_c = 0.06, and ρvk = 0.18 h⁻². The circles are experimental values from Ref. 48.*

events and by ξ(t) the conversion rate at time t one gets (46)

$$1-\xi(t)=e^{-2\rho v(e^{-kt}+kt-1)} \qquad (10)$$

where 1/k is the activation time for such a site $(v\rho)^{-1}$ is the propagation time between two sites which can be activated and critically depends on the contrains along the chain. These contrains result form the repeat unit mismatch ; they are more important for $\xi>\xi_c$ where $\xi_c \simeq 10\%$ which is the critical conversion rate where the autocalytic effect sets in. Let us assume that the critical conversion rate ξ_c at a given temperature T is reached at time t_c at that ρv has the values $\lambda 0$ and λ_∞ before and after the time t_c respectively. Then one gets from (10) with simple probabilitistic arguments

$$1-\xi(t)=e^{-\lambda_0 t} e^{-\lambda_\infty k(t-t_c)^2\ Y(t-t_c)} \qquad (11)$$

whereY(τ) is the Heaviside step function. As can be seen on Figure 7 this reproduces remarkably well the experimental results (48). More detailed considerations using transition state theory allow to obtain a rough estimate of t_c. We insist however that the above approach is essentially applicable to the PTS-polydiacetylene or similar ones and drastic deviations may occur for other compounds.

Conclusion

The linear and nonlinear optical properties of one-dimensional conjugated polymers contain a wealth of information closely related to the structure and dynamics of the π-electron distribution and to their interaction with the lattice distorsions. The existing values of the nonlinear susceptibilities indicate that these materials are strong candidates for nonlinear optical devices in different applications. However their time response may be limited by the diffusion time of intrinsic conjugation defects and the electron-phonon coupling. Since these defects arise from competition of resonant chemical structures the possible remedy is to control this competition without affecting the delocalization. The understanding of the polymerisation process is consequently essential.

Literature Cited

1. C. Flytzanis, in Nonlinear Behaviour of Molecules Atoms and Ions in Electrics, Magnetics or Electromagnetic Fields Ed. Trojanowski, Elsevier, Amsterdam, 1979, p. 185
2. G. Wegner, in Chemistry and Physics of One-dimensional Metals, ed. H.J. Keller, Plenum, N.Y., 1977, p.297
3. C. Sauteret, J.P. Hermann, R. Frey, F. Pradère, J. Ducuing, R.R. Chance and R.H. Baughman, Phys.Rev.Lett., 36, 956 (1976)
4. G.P. Agrawal, C. Flytzanis, Chem.Phys.Lett., 44, 366 (1976)
5. R.H. Baughman, J.Appl.Phys., 43, 4362 (1972)
6. D.L. Greenaway, G. Harbeke, F. Bassani and E. Tosatti, Phys. Rev., 178, 1340 (1969)
7. D. Reimer, M. Bässler, H. Hesse and G. Weiser, Phys.Stat.Sol., 73b, 709 (1976)
8. D. Bloor, F.H. Preston and D.J. Ando, Chem.Phys.Lett., 38, 33 (1976)
9. G. Wegner, Die Makromolekulare Chemie, 145, 85 (1971)
10. C. Bubeck, H. Sixl, H.C. Wolf, Chem.Phys., 32, 231 (1978) ibid 48, 269 (1980) ; ibid 50, 273 (1980)
11. C. Flytzanis, Polymer Preprints, vol 23, n°2, 153 (1982)
12. C. Flytzanis, in Treatise of Quantum Electronics vol.Ia, ed. H. Rabin and C.L. Tang, Academic Press, N.Y., 1975, p.19
13. S.K. Kurtz, ibid p.210
14. J.P. Hermann and J. Ducuing, J.Appl.Phys., 45, 5100 (1974)
15. J.L. Oudar, J.Chem.Phys., 67, 446 (1977)
16. B.F. Levine and C.G. Bethea, J.Chem.Phys., 65, 2429 (1976)
17. A. Dulcic and C. Flytzanis, Opt.Comm., 25, 402 (1978)
18. J. Harrison, Electronc Structure and the Properties of Solids The Physics of the Chemical Bond, W.H. Freeman and Co, San Francisco, 1980.
19. H. Jones, Theory of Brillouin Zones and Electronic States in Crystals, North-Holland, Amsterdam, 1960
20. V.N. Genkin and P.M. Mednis, Sov.Phys.JETP, 27, 609 (1968) (Russ.ed. 54, 1137 (1968))
21. C. Cojan, G.P. Agrawal, C. Flytzanis, Phys.Rev., B15, 909 (1977)
22. G.P. Agrawal, C. Cojan and C. Flytzanis, Phys.Rev., B17, 776 (1978)
23. J.N. Murrell, the Theory of the Electronic Spectra of Organic Molecules, Methuen, London (1963)
24. M. Cardona and F.H. Pollack, in Optoelectronic Materials, ed. G.A. Albers, Plenum, N.Y., 1971
25. A. Garito and K. Singer, Laser Focus February 1982, p. 59
26. H.G. Häfele, H. Wacherning, C. Islinger, R. Grisar, and R. Nitsche, Phys.Stat.Sol.,42, 531 (1970)
27. G.P. Agrawal, C. Cojan and C.Flytzanis, Phys.Rev.Lett., 38, 711 (1977)
28. A.A. Ovchinnikov, Soviet.Phys., Uspekki, 15, 575 (1973)
29. F. Kaizar, J. Messier , J. Zyss and I. Ledoux, Opt.Comm., 45, 133 (1983)

30. M. Lequime and J.P. Hermann, Chem.Phys., 26, 431 (1977)
31. M.I. Shand, R.R. Chance and R. Silbey, Chem.Phys.Lett., 64, 448 (1979)
32. C. Minot, C. Flytzanis, Chem.Phys.Lett., 68, 501 (1979)
33. R. Peierls, Quantum Theory of Solids, Clarencon Press, Oxford 1955
34. J.A. Pople and S.H. Wamsley, Mol.Phys., 5, 15 (1962)
35. S.A. Brazovskii, J.E.T.P. Lett., 28, 606 (1978)
36. W.P. Su, J.R. Schrieffer and A.J. Heeger, Phys.Rev., B22, 2099 (1980)
37. C. Flytzanis, to be published
38. J.P. Hermann and P.W. Smith, XIth Inter.Quantum el. Conference 1980, Digest of Technical Papers, IEEE, p. 656 (1980)
39. W. Voigt, Phys.Zeitschr., 17, 287 and 307 (1916) ibid., 18, 59 (1917)
40. H. Sixl, Preprint
41. R.H. Baughman, J.Chem.Phys., 68, 3110 (1978)
42. C. Flytzanis, to be published
43. R.R. Dogonadze, A.M. Kuznetsov, M.A. Vorotyntsev, Phys.Stat. Sol., 54b, 125 (1972)
44. R. Kubo and Y. Toyozawa, Progr.Theor., 13, 160 (1955)
45. C. Flytzanis, C.L. Tang, Phys.Rev.Lett., 45, 441 (1980)
46. T.H.K. Barron and E.A. Boucher, preprint
47. C. Minot, thèse de 3ème cycle, Université de Paris-Sud, 1980
48. D. Bloor, L. Koski, G.C. Stevens, F.H. Preston and D.J. Ando J. of Mat.Sc. 10, 1678 (1975)

RECEIVED July 29, 1983

9

Nonlinear Optical Properties of Polydiacetylenes

M. L. SHAND and R. R. CHANCE

Allied Corporation, Corporate Research and Technology, Morristown, NJ 07960

The nonlinear optical properties of novel, soluble polydiacetylenes are reviewed and discussed. The nonlinear optical properties are determined using resonance Raman scattering, coherent anti-stokes Raman scattering and coherent stokes Raman scattering. The two-photon polarizability is found to be very large in these materials. General aspects of the third-order susceptibility of these materials are also discussed.

The linear and nonlinear optical properties of polydiacetylenes have received considerable attention recently (1-3). This interest is attributable to the fully-conjugated backbone structure

A B

which leads to essentially one dimensional electronic properties. The acetylene mesomer (A) is the lowest-energy conformation for the polymer, and numerous x-ray structures have yielded bonding sequences which are very close to A above (4). There is some evidence for the higher-energy butatriene mesomer (5, 6) in systems, such as the ones we are considering here, where backbone strain is introduced due to intramolecular interactions of the substituent groups (R).

0097-6156/83/0233-0187$07.25/0

Extensive electron delocalization along the chain direction leads to an electronic transition energy for one-photon absorption (E_0) of typically 15 000-16 000 cm^{-1} for an unstrained backbone (7). This value is very close to that of polyacetylene (or polyene) $(HC=CH)_n$ which has $E_0 \sim 14\ 000\ cm^{-1}$ (8). For both polymers, E_0 varies with chain length (or conjugation length) in much the same manner (9). This analogy to the polyene system leads to the expectation that the polydiacetylenes might have unusually large third-order susceptibilities (χ) as had been demonstrated for polyenes of varying length. This expectation was borne out by the work of Sauteret et al. (1) who found χ values for polydiacetylene crystals comparable to those of inorganic semiconductors. Their results suggested attractive possibilities for these one-dimensional materials in nonlinear optical devices. Applications in parametric oscillators which utilize three-wave mixing were particularly attractive. However, three-wave-mixing experiments were unsuccessful due to phase-matching problems (10), as well as nonlinear absorption problems which Lequime and Hermann (11) attributed to absorption by photogenerated defects.

Recently, polydiacetylenes have been discovered which are soluble in common organic solvents (12). These polymers have urethane substituent groups of the form

$$-(CH_2)_m O\overset{\overset{\displaystyle O}{\|}}{C}-NH-X$$

where m=3 or 4. In the work discussed here X is $-COOC_4H_9$; these materials are referred to as mBCMU (BCMU = butoxycarbonylmethylurethane). The most interesting feature of these polymers is that the visible absorption spectra can be varied in a controlled manner by solvent variation (see Fig. 1). This phenomenon is due to the fact that the degree of conjugation (or planarity) of the chains in solution is controlled by intramolecular hydrogen bonding of the C=O and N-H functionalities on adjacent urethane substituent groups. In good solvents, such as $CHCl_3$, about one out of every four polymer repeat units (r.u.) is not hydrogen bonded, which leads to a conjugation length of ~4 repeat units (r.u.) and a large blue shift in optical absorption ($E_0 \sim 21\ 000\ cm^{-1}$) (12). In poor solvents, such as $CHCl_3$-hexane mixtures, essentially every polymer unit is hydrogen bonded, which leads to $E_0 \sim 16\ 000\ cm^{-1}$ for 3BCMU and $E_0 \sim 19\ 000\ cm^{-1}$ for 4BCMU. The blue shift of the latter compared to the usual crystal values is attributed to strain (12, 13).

This paper will review the linear and nonlinear optical properties of polydiacetylenes with an emphasis on our work with the nBCMU polymers. The following section will discuss material

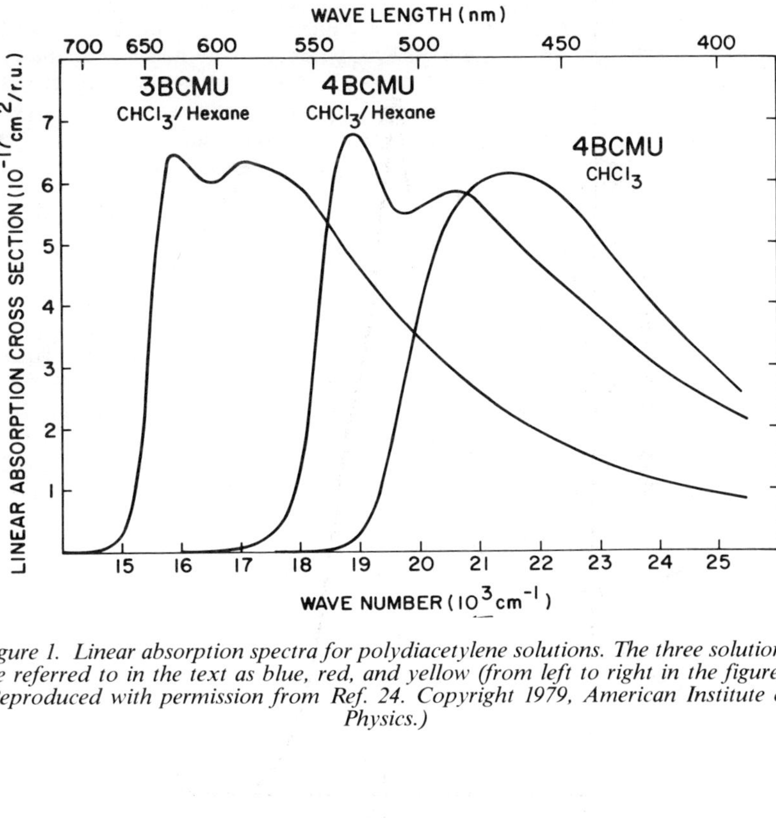

Figure 1. Linear absorption spectra for polydiacetylene solutions. The three solutions are referred to in the text as blue, red, and yellow (from left to right in the figure). (Reproduced with permission from Ref. 24. Copyright 1979, American Institute of Physics.)

properties including the conformational transition. The next sections include the linear optical properties of polydiacetylenes and how they relate to the material properties, the nonlinear optical properties and, finally, the present understanding of these properties.

Material Properties and Linear Optical Properties

The solid-state polymerization of diacetylenes is an example of a lattice-controlled solid-state reaction. Polydiacetylenes are synthesized via a 1,4-addition reaction of monomer crystals of the form R-C≡C-C≡C-R. The polymer backbone has a planar, fully conjugated structure. The electronic structure is essentially one dimensional with a lowest-energy optical transition of typically 16 000 cm^{-1}. The polydiacetylenes are unique among organic polymers in that they may be obtained as large-dimension single crystals.

Solutions of CMU polymers show dramatic color changes when the solvent/nonsolvent ratio is varied. The color changes have their origins in a conformational transition which can be viewed as a one-dimensional crystallization of a single chain in the polymer solution. 3BCMU is quite soluble in chloroform (>5% by weight or >0.1 r.u. mole/liter). The 3BCMU polymer solution, at sufficient dilution, is yellow in appearance (E_0~21 000 cm^{-1}). On addition of hexane, a dramatic reversible color transition to a dark blue solution occurs (E_0~16 000 cm^{-1}); subsequent addition of hexane results in the precipitation of the polymer as a blue solid. The blue solution is a true solution; it is quite stable and does not precipitate on standing or during centrifugation. Since the color transition is independent of polymer concentration, it is clearly a single-chain phenomenon. Completely analogous results are obtained for 4BCMU except the transition is from yellow (E_0 = 21 000 cm^{-1}) to red (E_0 = 19 000 cm^{-1}).

Absorption spectra for 3BCMU are shown in Figure 2. The broad structureless spectra of the yellow solutions (X_{CHCl_3}= 0.87, 1.0) are not unexpected; in fact, the behavior of the longer-chain polyenes is quite analogous (14). The optical properties of a conjugated system are determined not by the chain length but by the conjugation length ℓ, i.e., the length over which planarity of the conjugated units is maintained without interruption. Rotation about the single bond in the polydiacetylene backbone would interrupt conjugation at a cost of ~ 1 kcal/mole in resonance energy (11) - a cost easily suplied by the increased entropy of the chain and the bulky, flexible side groups. Such rotations would be facilitated by the interaction of the chloroform solvent with the ester and urethane units of the R group, an interaction

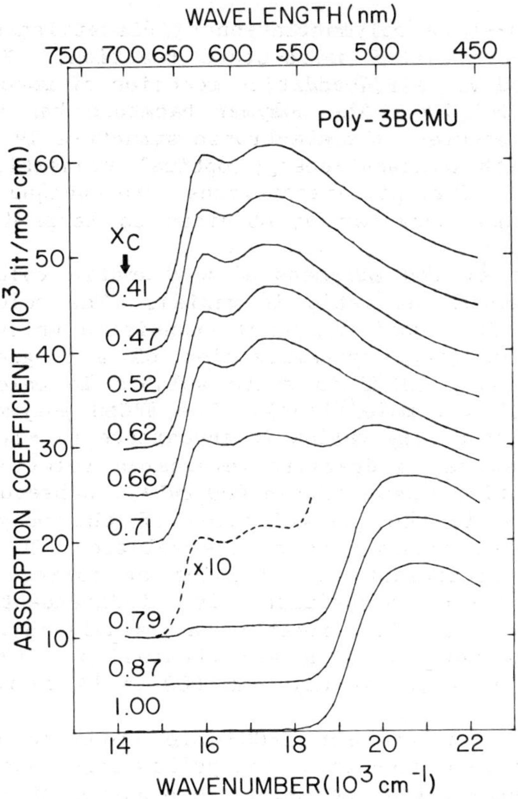

Figure 2. Visible absorption spectra of solutions of 3BCMU for various mole fractions of chloroform in hexane. The spectra are progressively offset by multiples of 5000 L/mol-cm. The polymer concentration is 1.94 × 10^{-5} mol/L.

which is undoubtedly a major factor in the high solubility of the CMU polymers. Based on the analogous polyene behavior, we can estimate that the absorption peak corresponds to $\ell \approx 7$ r.u. (or 14 conjugated multiple bonds) in the yellow solution (9, 16). Thus, in the yellow solution the backbone has acquired a basically nonplanar conformation.

As the nonsolvent hexane is added the solvent-polymer interaction is decreased - a process which would eventually lead to precipitation. In this case, however, a conformational transition occurs from the nonplanar polymer structure of the yellow solution to a planar structure in the blue solution, giving rise to the 16 000 cm^{-1} optical transition. This optical transition energy is essentially the infinite conjugation length limit, since it is very close to that typically observed for polydiacetylene crystals.(7) Clearly, the conjugation lengths are quite long; ℓ is estimated to be >30 r.u. for the blue solution, based on comparison to polyene results (16). Peak absorption coefficients for the blue solution are in good agreement with the polyene results (17) and with results for polydiacetylene crystals (18, 19).

It is important to point out that in Figure 2, though intermediate spectra (X_{CHCl_3}=0.79, 0.71) are observed in the transition, there is no indication of intermediate electronic species. The observed changes are discontinuous (in terms of spectral shifts) so that the 16 000 cm^{-1} optical transition is already evident at the earliest stages of the color change. This suggests some cooperative behavior as would be expected with nucleation and growth in the proposed one-dimensional crystallization process.

The stabilization of the planar conformation in these materials is due primarily to intramolecular hydrogen bonding between the N-H and C=O (urethane) groups on adjacent R groups. The resulting polymer conformation is shown schematically in Figure 3. This explanation of the color transition is clearly evident from infrared spectra (12), which will not be discussed here.

Completley analogous results have been obtained for poly-4BCMU, though changes in the visible and infrared spectra are considerably less dramatic (5). For example, the frequency shift in the visible absorption spectrum is ~2400 cm^{-1} for poly-4BCMU compared to ~4400 cm^{-1} for poly-3BCMU (Fig. 2). Molecular models sugest that the hydrogen bonds are more easily formed in the 3BCMU case.

Nonlinear Optical Properties

The aspects of nonlinear optics as applied to polydiacetylenes which will be discussed in this section include

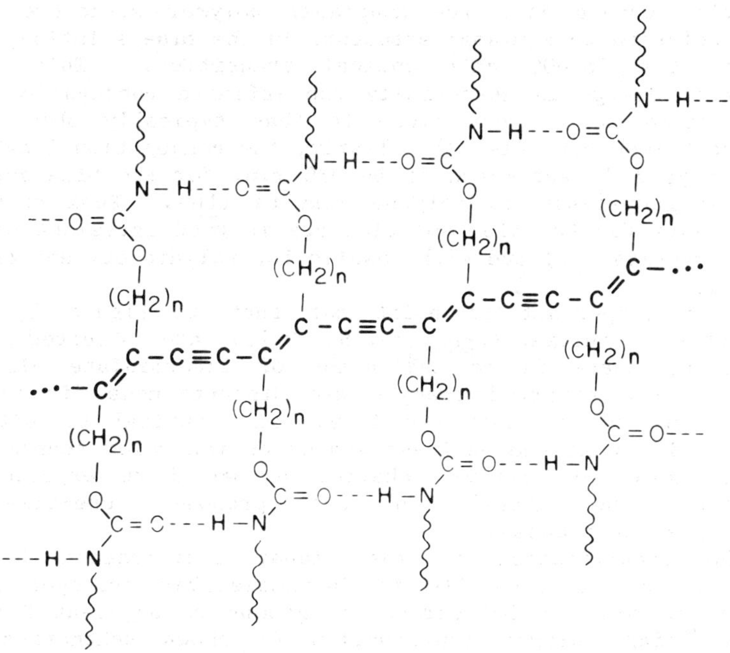

Figure 3. CMU polymer in planar (hydrogen bonded) configuration. The structure shown is that suggested by molecular models of 3BCMU where n = 3 *and* $Z = -C_4H_9$, *i.e. the wavy line in the figure at the end of the substituent group is* $-CH_2COOC_4H_9$. *(Reproduced with permission from Ref. 12. Copyright 1979, American Institute of Physics.)*

Raman scattering (RS) and resonance Raman scattering (RRS), which show conclusively that the broad absorption line shape is due to a distribution of chains of varying conjugation lengths rather than a homgeneously broadened line width, and three wave mixing (TWM) which shows that the large $\chi^{(3)}$ in polydiacetylenes is dominated by a strongly allowed two-photon transition.

We begin with a brief mention of the experiments of Sauteret et al. (1) who measured third harmonic generation in polydiacetylene crystals and found $\chi^{(3)}$ values comparable to those of inorganic semiconductors. They observed a 1000 fold enhancement of $\chi^{(3)}$ in the polymer compared to the monomer and an anisotropy of at least 100, with $\chi^{(3)}$ largest in the polymer chain direction. $\chi^{(3)}$ was measured for two polymers and at two incident frequencies, 3820 cm^{-1} and 5290 cm^{-1}. The largest value of $\chi^{(3)}$ (~ 10^{-10} esu) was obtained for the polydiacetylene PTS [R is $-CH_2SO_3C_6H_4CH_3$] at 5290 cm^{-1}. The hyperpolarizability ($\gamma=\chi/N$ ignoring local field effects) is then about 5 x 10^{-32} esu in the polymer chain direction. Note that 3ω (15 870 cm^{-1}) is quite near the band edge of PTS (16 200 cm^{-1}) and resonance effects are undoubtedly important. In fact, with 3820 cm^{-1} excitation, $\chi^{(3)}$ is found to be about five times less. Similarly in another polymer with a higher band gap, $\chi^{(3)}$ values were significantly lower. This work opened up attractive possibilities for polydiacetylene as nonlinear optical materials (1, 20).

Raman Scattering. The RS and RRS (21, 22) measurements actually were done after the TWM (23-25), but will be described first because the TWM analysis used important assumptions later confirmed by the RRS work. The RRS process enhances the ordinary RS process when the incident light energy (ω_i) or the scattered light energy (ω_s) is near resonance with electronic states in the scattering medium. In a sample with several components, RRS is selective, enhancing only those chromophores which have electronic states near ω_i or ω_s. In this study ω_i is shifted throughout the absorption band of polydiacetylene solutions to test the hypothesis that these solutions are composed of a number of chromophoric components each of which has different electronic states contributing to different regions of the absorption band. As ω_i is shifted the RS from different components will be enhanced and the observed vibrational frequencies will be those of the photoselected component.

The electronic properties and thus the absorption bands of a polymer depend on the conjugation length ℓ. The absorption energy decreases as ℓ increases reaching a limiting value of ~ 16 000 cm^{-1} as ℓ approaches infinity. If the polymer solution consists of a distribution of conjugation lengths (i.e. chromophores), photoselection of the different chromophores within the inhomogeneously broadened absorption band should be possible with RRS. The Raman frequencies for backbone carbons are expected to vary with ℓ for basically the same reason that optical absorption varies with ℓ, i.e., electron delocalization. In particular, as electron delocalization increases with increasing ℓ, the C=C bonds and the C≡C bonds become weaker and the Raman active vibrational frequencies associated primarily with these bonds, $\nu^{C=C}$ and $\nu^{C \equiv C}$, will decrease. As usual, in the following, the 3BCMU in $CHCl_3$-hexane is referred to as the "blue solution" and 3BCMU in $CHCL_3$ is referred to as the "yellow solution".

An example of a RRS spectrum from the yellow solution with λ_i=488 nm (ω_i=20 492 cm^{-1}) is shown in Fig. 4. Also shown are spectra in the $\nu^{C=C}$ regions. The Raman modes of interest are the 667 cm^{-1} $CHCl_3$ peak which is used for polymer mode normalization, $\nu^{C=C}$ at 1520 cm^{-1} and $\nu^{C \equiv C}$ at 2120 cm^{-1}. Analysis of the peaks in the yellow solution is straightforward because the Raman bands do not overlap.

Raman spectra of $\nu^{C=C}$ in the blue solution are complicated because of hexane lines at 1450 cm^{-1} and by modes which do not shift with ω_i. Peak intensities can be inferred by comparison to the nonresonant hexane signal near 1450 cm^{-1} which shows only the usual nonresonant $\omega_i{}^4$ variation. A line at 1460 cm^{-1} is present which does not shift with incident frequency. $\nu^{C=C}$ for 3BCMU polymer single crystal ($\ell \to \infty$) is 1460 cm^{-1}. Thus, for λ_i greater than approximately 540 nm the RS indicates an absorption spectrum with chains having a single $\nu^{C=C}$ Raman frequency - the frequency value indicating an effectively infinite conjugation length, as does the E_o value.

The variation of $\nu^{C=C}$ is shown in Fig. 5 for both the yellow and blue solutions. For comparison the absorption spectrum of each solution is also shown in Fig. 5. In the yellow solution at small ω_i all chromophores are in preresonant conditions and all contribute to the Raman line, which is broadened (width larger than 30 cm^{-1}) and peaks at an average value of ≃1520 cm^{-1}. At lowest energy and farthest from resonance the Raman line is asymmetrically broadened as expected (Fig 4). As ω_i is increased, approaching the absorption band, chromophores with largest ℓ reach resonant conditions, their contribution to the Raman line dominates, and the peak frequency decreases to the value characteristic of these chains. As ω_i is increased further through the absorption band, chrompohores of successively smaller average ℓ values are in resonance (photoselected) and dominate the spectrum. Thus, the peak frequency increases with ω_i. In all RRS spectra the linewidth is nearly constant at 26 cm^{-1}, decreasing only slightly with increasing ω_i. The blue solution results are similar in the high energy region though the short chain component to these solutions is clearly less than in the yellow solution.

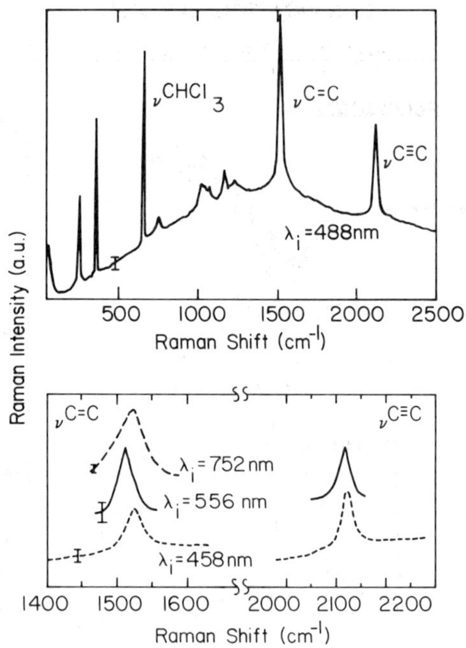

Figure 4. Raman spectra of the yellow solution. Key: top, entire spectrum with λ_i = 488 nm; bottom, spectra of the $\nu^{C=C}$ and $\nu^{C\equiv C}$ regions with λ_i = 458 nm (---) and λ_i = 556 nm (—) and $\nu^{C=C}$ regions with λ_i = 752 nm (--). (Reproduced with permission from Ref. 21. Copyright 1982, American Institute of Physics.)

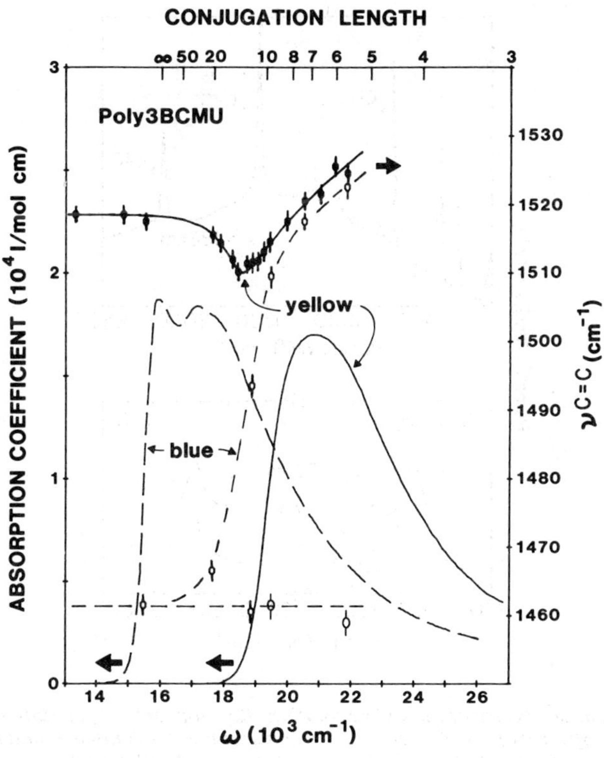

Figure 5. Absorption spectra and Raman data for 3BCMU yellow and blue solutions. The data points indicate the ω_i dependence of $\nu^{C=C}$. The conjugation length scale is an estimate of the length of a chromophore with absorption frequency ω. (Reproduced with permission from Ref. 21. Copyright 1982, American Institute of Physics.)

A model has been given (25) which attempts to quantify the qualitative interpretation given above. As previuosly discussed the results are described in terms of photoselection of chromophores (i.e., conjugated units of length ℓ r.u.). The first step in this analysis is to model the distribution of chain lengths using the absorption spectrum, which is shown in Fig. 6 (solid curve) for the yellow solution. The absorption peak for this solution (21 000 cm^{-1}) corresponds to $\ell \simeq 6$-7 units. However, infrared studies of side group interactions clearly indicate an average ℓ of 3-4 units. We attribute this difference to an increase in the electronic linewidth Γ_ℓ with decreasing ℓ. This increase can be expected due to, for example, lifetime shortening because of intrachain, nonradiative energy transfer from short to long chain segments. A model based on these ideas and a Gaussian distribution of chain lengths gives the results shown in Fig. 6. There are two free parameters in this fit.

The Raman modes $\nu^{C=C}$ and $\nu^{C\equiv C}$ depend on chain length in a manner which parallels the absorption energy dependence on chain length (25). The contribution of each chromophore to the observed Raman peak is determined by the RS intensity for incident light at energy ω_i. The RS peak position is then a weighted average of all the Raman modes of all chromophores. The fit to the data with the model (same as in Fig. 6) is shown in Fig. 7 for $\nu^{C=C}$ and for $\nu^{C\equiv C}$ in the yellow solution. In both cases the experimental data are quite well represented considering that there is only one parameter in the fit. The model also adequately describes the Raman excitation profile. The good agreement between the model and the experimental results confirms the basic premise of the model that the individual polymer chains in solution contain a distribution of chain lengths which determine the absorption characteristics of the solutions.

We should note in passing that the above work on Raman photoselection is directly relevant to the current controversy regarding similar solid-state experiments on polyacetylene, $-(HC{=}CH)_n-$. The linear optical properties of polyacetylene and polydiacetylene are very similar and the close analogy between the two systems has been emphasized in previous work (9). Kuzmany *et al.* (26) observe effects in Raman spectra of polyacetylene which are quite similar to those seen in our blue solutions and apply basically the same interpretation, i.e., conjugation length dispersion. Mele (27) and Lauchlan et al. (28) argue against this interpretation and suggest a model based on hot luminescence. The latter model is not consistent with our yellow solution data. Therefore, the internal consistency between our yellow and blue solution data and the similarity of the latter to the polyacetylene case argues heavily in favor of the conjugation-length dispersion model for polyacetylene.

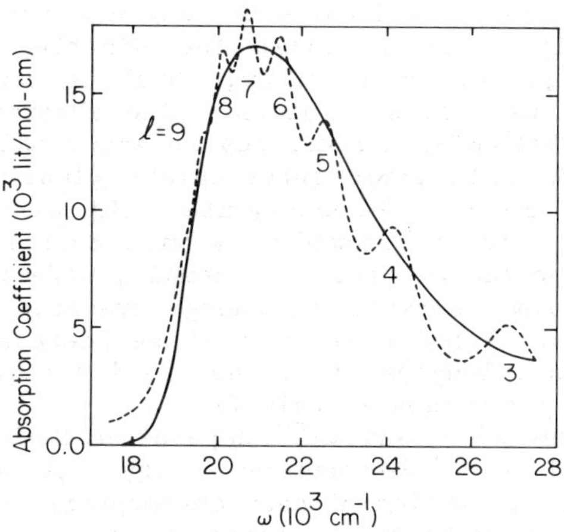

Figure 6. Comparison of experimental (—) and theoretical (---) absorption spectra for the yellow solution. Peaks in the theoretical curve are due to contributions from chains of length l and are labeled with the appropriate l. (Reproduced with permission from Ref. 21. Copyright 1982, American Institute of Physics.)

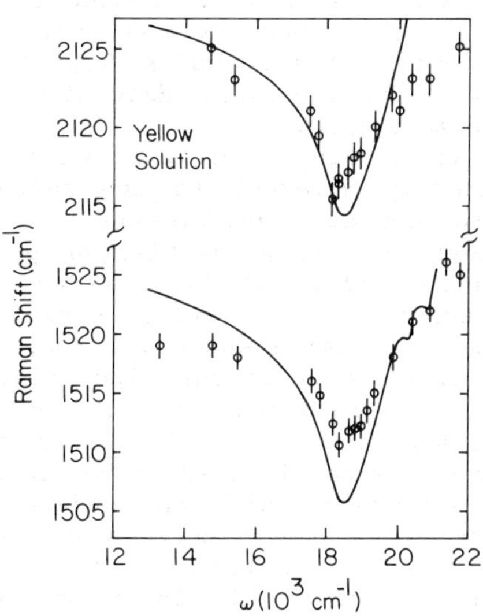

Figure 7. Experiment and theory for the excitation energy dependence of $\nu^{C=C}$ and $\nu^{C\equiv C}$. (Reproduced with permission from Ref. 21. Copyright 1982, American Institute of Physics.)

In the next section, we will discuss three wave mixing experiments (two-photon absorption) in these solutions. The Raman results provide important background information for two reasons: 1) the conjugation-length dispersion model is confirmed by the Raman data and suggests a similar dispersion in the energies of two-photon accessible states and 2) the Raman results demonstarte that, over the pre-resonance frequency range of our three wave mixing experiments, the laser "sees" the ensemble average of the conjugation length distribution. The latter is reflected in the average value of $\nu^{C=C}$ and $\nu^{C\equiv C}$ observed up to $\omega_i \sim 17\ 500\ cm^{-1}$ in the yellow solution.

Three Wave Mixing. A number of resonances are possible in the third-order susceptibility, $\chi^{(3)}$. These resonances may be due to phonons or vibronic excitation such as in CARS (Coherent Anti-Stokes Raman Scattering), electronic states accessible to one or two photons (typically in third harmonic generation), or any other excitation in the nonlinear material. In general, $\chi^{(3)}$ represents "four-photon" or "four-wave" processes which have been called four-wave mixing (FWM). The process described here in which only two input beams are present (one beam is utilized twice in each process) has been called three-wave mixing (TWM) (29) as well as FWM. The traditional TWM will be used in this paper.

TWM is accomplished by focusing two laser beams with frequencies ω_1 and ω_2 in the polymer solution. Incident laser beams are generated by two dye lasers pumped by a single N_2 laser. Wavelengths for the incident beams ranged from 760 nm to 590 nm (or to the region of large absorption). The beams are focused into the sample with a 15-cm focal length lens. The crossing angle between the two beams is adjusted for phase matching (approximately 4° outside the sample cell). The sample solutions are held in a 1-cm cuvette. The beam generated at $\omega_3 = 2\omega_1 - \omega_2$ is filtered with a double monochromator and its intensity measured with a photomultiplier and boxcar electronics. Our analysis utilizes the measurement of the ω_3 intensity as a function of polymer concentration to determine the real and imaginary parts of the two-photon hyperpolarizability (24, 30). The ω_3 intensity of the solvent serves as an internal intensity reference. The first discussion of TWM experiments will be of the two photon absorptions. Finally, the Raman resonances in $\chi^{(3)}$ will be discussed.

In order to extend the range of $2\omega_1$ possible with a limited range of laser excitation, both CARS (Coherent Anti-Stokes Raman Scattering) and CSRS (Coherent Stokes Raman Scattering) are used. In both cases $\omega_3 = 2\omega_1 - \omega_2$. In the CARS mode $\omega_3 > \omega_1 > \omega_2$; in the CSRS mode $\omega_2 > \omega_1 > \omega_3$. One-photon resonance effects are the same in both cases as described later. Phase matching is also the same in both cases with $\overline{k_3} = 2\overline{k_1} - \overline{k_2}$.

Linear absorption spectra for the three solutions considered herein are given in Fig. 1. The concentration range available in the yellow solution is 0-120 x 10^{17} r.u./cm^3. For the other solutions an upper limit is imposed on the concentration range due to polymer precipitation: ~4 x 10^{17} r.u./cm^3 for the red soution and ~9 x 10^{17} r.u./cm^3 for the blue solution.

The two-photon hyperpolarizability is determined from the variation of the ω_3 intensity with polymer concentration. This concentration dependence is determined most accurately for the yellow solution, since the concentration range is largest in that case. In addition, the low-energy tail of the linear absorption (which corresponds to the high-frequency cut off of our experimental range) is at higher energies in the yellow solution than in the blue or red solutions. Therefore, the yellow solutions yield the more extensive data sets and these results are presented and discussed in detail here.

Six samples of yellow solution were prepared with various concentration up to 1.2 x 10^{19} r.u./cm^3. At higher concentrations and in regions of linear absorption the phase-matching conditions depend on concentration. These higher concentrations are not used in order to avoid changing the phase-matching angle during a series of measurements. The third-order nonlinear mixing is determined in both the CARS and CSRS geometries. $\omega_1 - \omega_2$ is tuned away from any Raman resonances at about 1300 cm^{-1}. The output intensity is measured for each sample and for the solvent without polymer (pure chloroform for the yellow solution). The total susceptibility is

$$\chi_{tot} = \chi_s + \chi_p + \chi_T = \chi_s + N\gamma_p + N(\gamma'_T + i\gamma''_T) \qquad (1)$$

where χ_s is the solvent susceptibility (assumed to be positive and real), χ_p is the polymer nonresonant susceptibility, χ_T is the two-photon susceptibility, N is the polymer concentration, γ_p is the polymer nonresonant hyperpolarizability, and $\gamma_T = \gamma'_T + i\gamma''_T$ is the two-photon hyperpolarizability. (Note that we take $\gamma = \chi/N$, ignoring local-field effects (24)). The ω_3 intensity I is proportional to $|\chi|^2$ so that

$$\frac{I^{s+p}}{I^s} = \frac{|\chi_{tot}|^2}{|\chi_s|^2} = 1 + \frac{2NR}{\chi_s} + \frac{N^2(R^2 + \gamma_T''^2)}{\chi_s^2} \qquad (2)$$

where I^{s+p} is the ω_3 intensity for the polymer solution, I^s is the ω_3 intensity for the pure solvent, and $R = \gamma_p + \gamma'_T$. It has been shown previously that the contribution of γ_p to R is very small over our frequency range (23-25). Therefore, R can be identified as the real part of the two-photon hyperpolarizability γ'_T. Note that, because of the interference of terms in $|\chi_{tot}|$, both the magnitude and sign of γ'_T is determined. Furthermore, we emphasize again that the solvent acts as an internal reference so that the results do not depend on measurement of absolute light intensities.

I^{s+p}/I^s as a function of N is shown in Fig. 8 for two values of ω_1. The closed and open circles are the data with $2\omega_1$ = 32 362 cm^{-1} and = 26 554 cm^{-1}, respectively. For the former, the intensity decreases with increasing N for small N, which indicates $\gamma'_T < 0$ for $2\omega_1$ = 32 362 cm^{-1}; for the latter, the intensity increases with increasing N indicating $\gamma'_T > 0$ for $2\omega_1$ = 26 554 cm^{-1}. Thus the energy of the two-photon state, determined by $\gamma'_T = 0$, is between these two values of $2\omega_1$. The solid lines in Fig. 8 are obtained from Eq.(2) with γ'_T/χ_s and γ''_T/χ_s determined from a linear regression analysis. The χ_s values are taken from Ref. 31: $\chi_s = 0.87 \times 10^{-14}$ esu for $CHCl_3$ and $\chi_s = 1.06 \times 10^{-14}$ esu for hexane. There are no two-photon resonances in the solvents in the frequency range of interest here (31). The resulting values of γ'_T and γ''_T, with 90% confidence limits expressed as error bars, are shown in Fig. 9. Each data pair is determined from the concentration dependence of I^{s+p}/I^s at one value of ω_1 in either a CARS or a CSRS configuration.

The solid lines in Fig. 9 are a least-squares fit of the real and imaginary parts to the theoretical form of γ_T (21):

$$\gamma_T \cong A/(\langle\omega_{gf}\rangle - 2\omega_1 - i\Gamma_{gf}) \quad (3)$$

where

$$A = \frac{3e^4\Phi f_{gi}f_{if}}{16hm^2\langle\omega_{gi}\rangle\langle\omega_{if}\rangle}\left(\frac{1}{\langle\omega_{gi}\rangle-\omega_1}\right)\left(\frac{1}{\langle\omega_{gi}\rangle-\omega_2}+\frac{1}{\langle\omega_{gi}\rangle-\omega_3}\right) \quad (4)$$

where ω_{gf} and Γ_{gf} are the energy and width of the state accessible to two-photon excitation (with only one such state assumed for an individual chromophore). The ground state is represented by g and the final (two-photon) state by f. The bracketed frequencies in Eqs. 3 and 4 represent averages over the ensemble of chromophores interacting with the photon fields. The terms in parentheses in Eq. 4 represent the one-photon resonance enhancement to the two-photon absorption term in χ_T. It is assumed that one state (labeled i) with energy ω_{gi} and width Γ_{gi} is responsible for these resonances. In general a summation over all electronic states and all conjugation lengths would be required. However, our lasers do not come in resonance with any of the available one photon states and we have demonstrated the appropriateness of this averaging

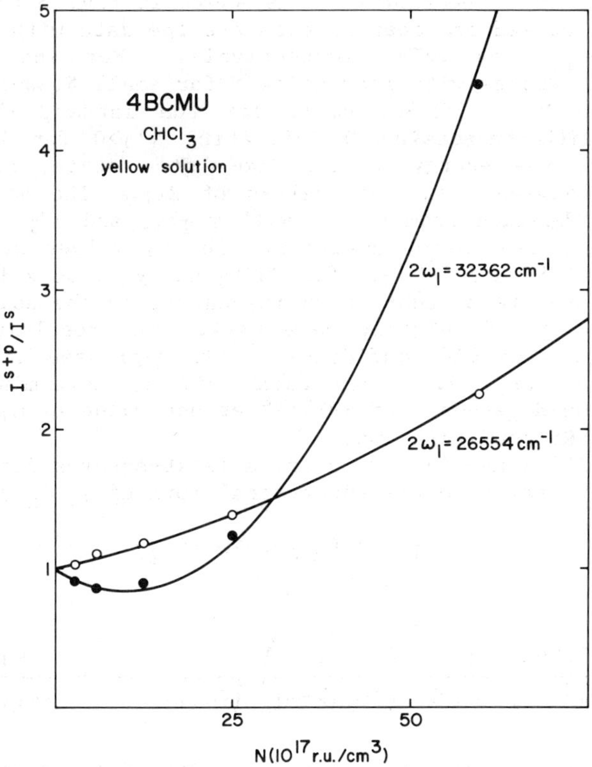

Figure 8. Variation in the normalized ω_3 output intensity with polymer concentration for the yellow solution at two incident ω_1 values. The solid lines represent theoretical fits to the data according to Equation 2. (Reproduced with permission from Ref. 25. Copyright 1980, American Institute of Physics.)

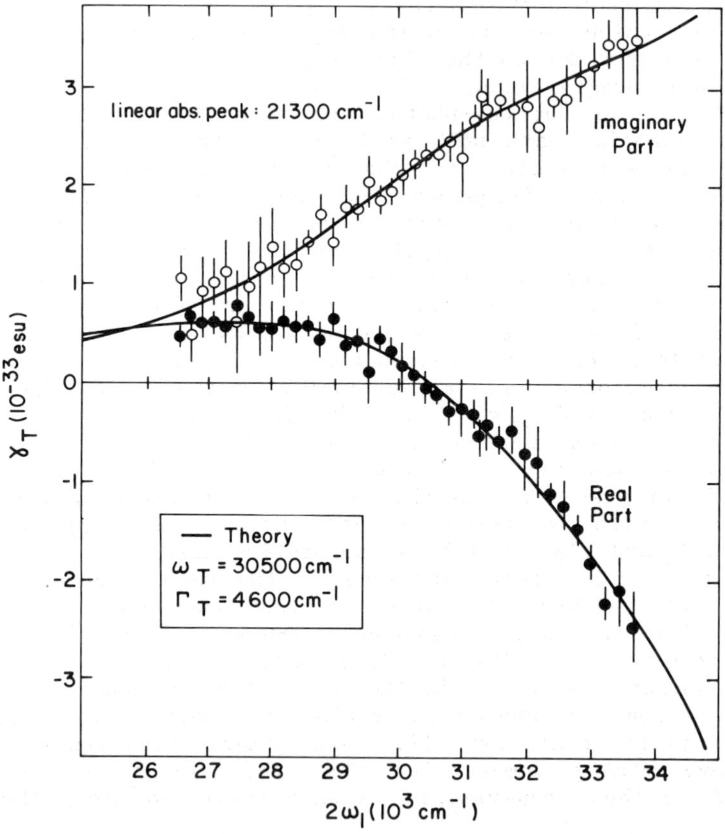

Figure 9. Variation of the real (●) and imaginary (○) parts of the two photon hyperpolarizability (γ_T) with two-photon energy ($2\omega_l$) for the yellow solution. The error bars represent 90% confidence limits. All data are taken in a 1111 geometry (all beams polarized ||). The solid lines are theoretical fits to the data according to Equation 3. (Reproduced with permission from Ref. 24. Copyright 1979, American Institute of Physics.)

procedure, as already mentioned, with our Raman results. (The width is ignored, since $\omega_{gi} - \omega >> \Gamma_{gi}$.) The factor Φ accounts for the fact that the TWM signal is an average over all molecular orientations in solution; Φ varies from 1 for an isotropic chromophore to 1/5 for a one-dimensional chromophore (our case). f_{jk} is the oscillator strength for the transition from j to k. The average over the ensemble of chromophores (i.e., chain lengths) has been approximated in Eq. (3) by a Lorentzian with a width Γ_{gf} which represents the inhomogeneous width of the two-photon resonance due to the distribution of chain lengths plus any vibrational progressions. The frequency $\langle\omega_{gi}\rangle$ and oscillator strength f_{gi} for the one-photon $g \rightarrow i$ transition are obtained from linear absorption data such as those in Fig. 1. Therefore, the only parameters varied in the fitting procedure are $\langle\omega_{gi}\rangle$, Γ_{gf} (the average frequency and inhomogeneous width of the distribution of states accessible by two-photon absorption), and f_{if} the oscillator strength for the transition from the intermediate state to the two-photon final state. Polarization measurements and theoretical calculations show that the state observed in linear absorption measurements is, as expected, the intermediate state in the two-photon resonance.

The two-photon state as determined by the fit to the yellow-solution data has $\langle\omega_{gf}\rangle$=30 500 cm^{-1}. The effect of the one-photon resonance enhancement on γ_T can be seen if we consider the form of the solid curves in Fig. 9 under conditions that no one-photon resonance is present. In this case, γ'_T would be a dispersive type curve with inversion symmetry around the γ'_T=0 point at 30 500 cm^{-1} and γ''_T would be an absorptive like curve centered at 30 500 cm^{-1}. The large increase in the magnitudes of γ'_T and γ''_T on the high-energy side of the spectrum is therefore attributed to one-photon resonance. The peak expected for γ''_T at 30 500 cm^{-1} is barely discernible as a broad shoulder, since it is almost completely obscured by the one-photon resonance.

The strong one-photon absorption is observed for polarizaion parallel to the chain axis (7). Our polarization studies of the three-wave mixing demonstrate that the two-photon absorption observed in these experiments is also polarized along the chain axis.

Similar measurements and analysis have been carried out for the red solution. In this case the concentration of polymer in the chloroform-hexane solvent is limited to less than 4 x 10^{17} r.u./cm^3. Therefore, the errors for γ'_T and γ''_T are larger than in the case of the yellow solution. In addition, the absorption band of the red solution is shifted to the red with respect to that of the yellow solution. Therefore, $2\omega_1$ can only be extended up to 32 000 cm^{-1}.

The blue solution, 3BCMU in chloroform and hexane, has

similar linear optical properties to the 4BCMU red solution but with a considerable red shift of the linear absorption spectrum (Fig. 1). Therefore, the accessible range of $2\omega_1$ is effectively limited to a single frequency (26 554 cm^{-1}), and we can only make estimates of $\langle\omega_{gf}\rangle$ and Γ_{gf}. A comparison of the concentration dependence for the yellow solution and the blue solution in Fig. 10 at the same ω_1 such that $\langle\omega_{gf}\rangle$(blue) $< 2\omega_1 <$ $\langle\omega_{gf}\rangle$(yellow) clearly shows the effect of the change of sign of γ'_T.

The values $\langle\omega_{gf}\rangle$ and Γ_{gf} determined from the fitting to $\gamma_T(2\omega_1)$ from the red and yellow solutions along with estimates for the blue solution are shown in Table I. The results for f_{gi} determined from linear absorption data and f_{if} determined from the γ_T data for all three solutions are also included in Table I. Note that these oscillator strengths correspond to the values appropriate for polarization parallel to the chain direction. Though the restricted data for the blue solution yields significant uncertainties in the $\langle\omega_{gf}\rangle$ and Γ_{gf} values, f_{if} for the blue solution is reasonably well determined nevertheless.

TABLE I. Energies, widths, and oscillator strengths for one- and two-photon absorption in polydiacetylene solutions (energies and widths in cm^{-1}).

Solution	$\langle\omega_{gi}\rangle$	f_{gi}[a]	$\langle\omega_{gf}\rangle$	Γ_{gf}	f_{if}[a]
Yellow	21 300	1.3	30 500	4600	2.7
Red	18 900	1.6	28 100	4000	2.5
Blue	15 900	1.4	~22 000	~3000	2.5

[a]Both f_{gi} and f_{if} represent oscillator strengths for polarization parallel to the chain direction.

The consistency of the oscillator strength for all three materials indicates that the only difference in the experimental results for each case are the resonance terms due to changes in ω_{gi} and ω_{gf}. The large oscillator strengths are an intrinsic property of the conjugated polymer which, together with one photon resonance effects, lead to the remarkably large two photon hyperpolarizabilities. The f_{gi} values are large, but not unexpectedly so; the f_{if} values given in Table I represent the first such determination to our knowledge. The magnitudes of both f_{gi} and f_{if} are understandable based on relatively simple theoretical considerations (25). We should note that the relative constancy in f_{gi} and f_{if} values for the different solutions suggests a lack of any conjugation length dependence - a result which can be expected based on polyene behavior (32).

It should be emphasized that the $\langle\omega_{gf}\rangle$ and Γ_{gf} values correspond to a Lorentzian description of the energy distribution

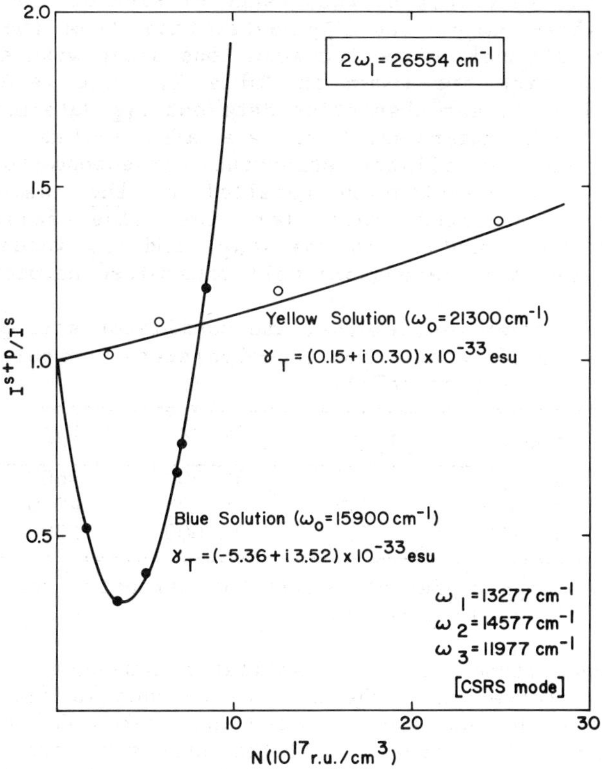

Figure 10. Variation of the normalized ω_3 output intensity (CSRS) vs. polymer concentration for the blue solution (●) and the yellow solution (○) at $2\omega_1 = 26{,}554\ cm^{-1}$. The solid lines are theoretical fits to the data: $\gamma_T = (0.15 + i0.30) \times 10^{-33}$ esu for the yellow solution and $\gamma_T = (-5.36 + i3.52) \times 10^{-33}$ esu for the blue solution. (Reproduced with permission from Ref. 25. Copyright 1982, American Institute of Physics.)

of two-photon accessible states. The inhomogeneous width Γ_{gf} is attributed to a distribution of conjunction lengths plus vibrational sidebands. It is reasonable to expect this width to be similar to or somewhat greater than the inhomogenous width of the one-photon transition (~ 2500 cm^{-1}; see Fig. 1). As a consequence of this fitting procedure, $\langle\omega_{gf}\rangle$ represents an average energy for the two-photon state and the electronic origin (0-0 two-photon transition energy) would probably be less than $\langle\omega_{gf}\rangle$ by at most a few thousand wave numbers, i.e., ~ Γ_{gf}. This effect would be most important for the red and blue solutions where a fairly well defined 0-0 transition is evident in the solution spectra of Fig. 1. Thus, for the blue solution a near coincidence of ω_{gi} and ω_{gf} is possible.

Based on these results, and on theoretical work which is in relatively good agreement with experiment (25), we suggest that the large two-photon hyperpolarizability is a general property of polydiacetylenes. The origin of the large γ_T values is a combination of large oscillator strengths and large one-photon resonance effects. Polydiacetylene crystals, which have ω_{gi} = 15 000-19 000 cm^{-1}, should also display large two-photon absorption TPA. This is an important point if these materials are to be considered for use in nonlinear optical devices based on the large susceptibilities measured with infrared lasers (1). There have been two previous measurements of TPA in polydiacetylene crystals. Reimer and Bässler (33) found a two-photon absorption coefficient, β, of ~ 5 x 10^{-48} cm^4-sec/photon-r.u. at 2ω=18 880 cm^{-1} for PTS (R in polydiacetylene structure is $-CH_2SO_3C_6H_4CH_3$; $\langle\omega_{gi}\rangle$=16 200 cm^{-1}). This value is large, though not remarkably so and is consistent with our values for γ''_T which is directly proportional to β.

Lequime and Hermann (11) measured TPA over an extended range in two polymers: PTS and TCDU [R is $-(CH_2)_4$ $OCONHC_6H_5$; $\langle\omega_{gi}\rangle$=18 500 cm^{-1}]. They utilized a 9400-cm^{-1} excitation beam and a low-intensity probe beam. Their results are in good agreement with the single-frequency result of Ref. 30· and indicate remarkably large β values, especially for PTS. They attributed this absorption to a two-step process involving defect formation and defect absorption. However, we are able to describe their data quite accurately based on TPA solution results (25). We conclude that the data of Lequime of Hermann, including the large difference they observed for two crystals with different ω_{gi} values, are easily explained by TPA without invoking defects. We further conclude that strong TPA is a fundamental property of the polydiacetylene backbone.

The large values of the hyperpolarizability in these molecules imply large susceptibilities for more concentrated solutions or solid-state films. This large third-order susceptibility can be utilized as a frequency shifter, even though χ has its origin in two-photon absorption (34). Linear absorption limits these devices to energies below 600 nm in 4BCMU, for example. This energy range is appropriate for extending the output of the alexandrite laser which produces line emission at 680 nm and tunable emission from 700 to 820 nm (35). TWM leads to generation of light with wavelengths as long as 1.1 µm. The efficiency of converting 680 nm radiation to the infrared for an oriented film of 4BCMU is calculated to be 1-10% depending on incident intensities. TPA limits the intensity of the incident beams which may be used.

Two experiments were attempted with higher density polymer materials. CSRS was observed in 4BCMU single crystal, ~69 µm thick, with ω_1 = 730 nm. Very little energy (~1.8 µJ) focused on the sample caused visible change; the sample turned red, presumably due to melting and conjugation length shortening. In a second experiment CARS was observed in a 4BCMU gel (36) with a density of ~10^{18} r.u./cm^3. The incident beam at ω_1 = 650 nm was not absorbed and no damage occurred. However, the CARS signal was strongly dependent on position of focusing in the sample due to inhomogeneities in the gel and was not as large as predicted from solution measurements. Therefore, these polymers are not at present practical parametric oscillators because of limitations of sample quality and, possibly, of TPA.

Resonances in $\chi^{(3)}$ also occur for Raman modes. Enhancement of the ω_3 output beam occurs when $\Delta\omega=\omega_1-\omega_2$ is near Raman active molecular vibrations as well as when $2\omega_1$ is near an electronic excitation. In this case there is an additional term to be added to $\chi^{(3)}$, Eq. (1), (23, 37):

$$\frac{NA}{\omega_p-\Delta\omega-i\Gamma_p} \qquad (7)$$

where N is the number density of polymer repeat units in r.u./cm^3, ω_p and Γ_p are the Raman mode frequency and width in the polymer. As $\Delta\omega$ is scanned over the polymer vibration at 1518 cm^{-1}, which is the C=C stretch mode, the third-order mixing signal varies as shown in Fig. 11. The solid curves are the experimental results. The dashed curves are least square fits of the data. The fits to the data use the results of the two-photon susceptibilities determined previously. The fit determines ω_p, Γ_p and A. Similar results are also obtained for Raman modes in the chloroform. The results are consistent with frequencies and widths determined by other means, and with the general model we have applied to these solutions.

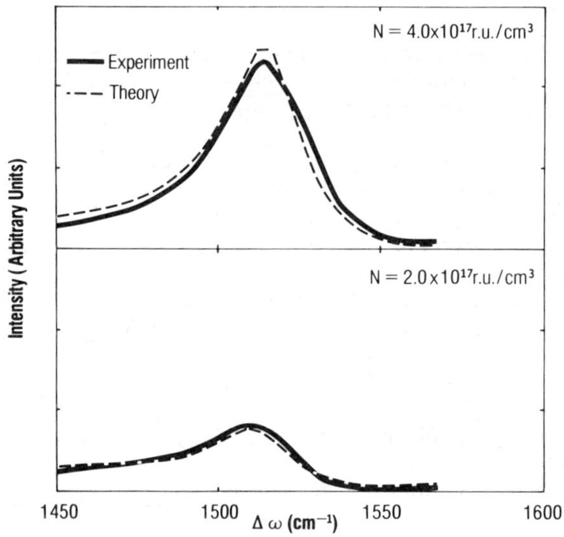

Figure 11. Third-order mixing (CARS) in a solution of 4 BCMU in 2/3 hexane and 1/3 chloroform for the frequency region of the polymer vibration. Key: —, experimental data; and ---, theoretical fit. (Reproduced with permission from Ref. 23. Copyright 1978, American Institute of Physics.)

Acknowledgments

Many people have contributed in various ways to this work. In particular we would like to thank M.A. Gilleo for many helpful discussions and R.H. Baughman, R. Silbey, C. Hogg, M. Schott, and M. LePostellec, for their collaborative input on various aspects of this work. We also acknowledge G.N. Patel for supplying materials and J.M. Sowa and M.C. Baker for valuable technical asistance.

Literature Cited

1. Sauteret, C.; Hermann, J.; Frey, R.; Pradère, F.; Ducing, J.; Baughman, R.H.; Chance, R.R. Phys. Rev. Lett. 1976, 36, 956.
2. Eckhardt, H.; Eckhardt, C.J.; Yee, K.C. J. Chem. Phys. 1979, 70, 5498.
3. Chance, R.R.; Baughman, R.H.; Müller, H.; Eckhardt, C.J. Phys Rev. Lett. 1977, 67, 3616.
4. See, for example, Bloore, D.; Fisher, D.A.; Batchelder, D.N.; Kennedy, R.; Cottle, A.C.; Lewis, W.F. Mol. Cryst. Liq. Cryst. 1979, 52, 83.
5. Enkelmann, V.; Lando, J.B. Acta Crystallogr. B 1978, 34, 2352.
6. Iqbal, Z.; Chance, R.R.; Baughman, R.H. J. Chem. Phys. 1977, 66, 5520.
7. Chance, R.R.; Patel, G.N.; Witt, J.D. J. Chem. Phys. 1979, 71, 206.
8. See, for example, Shirakawa, H.; Ito, T.; Ikeda, S. Polym. J. 1973, 4, 460.
9. Baughman, R.H.; Chance, R.R. J. Polym. Sci., Polym. Phys. Ed. 1976, 14, 2037.
10. Ducuing, J.; (private communication).
11. Lequime, M.; Hermann J., Chem. Phys. 1977, 26, 431.
12. Patel, G.N.; Chance, R.R.; Witt, J.D. J. Chem. Phys. 1979, 70, 4387.
13. Chance, R.R. Macromol. 1981, 13, 396.
14. Drabkin, I.A.; Tsaryuk, V.I.; Cherkasbin, M.I.; Kisilitsa, P.P.; Chauser, M.G.; Chigir, A.N.; Berlin, A.A. Vysokomol. Soedin. Ser. A 1968, 10, 1727.
15. Baughman, R.H.; Chance, R.R. J. App. Phys. 1976, 47, 4295.
16. Kuhn, H. Fortschr. Chem. Org. Naturst. 1958, 16, 169; ibid, 1959, 17, 404.
17. Feichtmayer, F.; Heilbronner, E.; Nürrenbach, A.; Pommer, H.; Schlag, J. Tetrahedron 1969, 25, 5383.
18. Reimer, B.; Bässler, H.; Hesse, J.; Weiser, G. Phys. Status Solidi B 1976, 73, 709.
19. Hood, R.; Müller, H.; Eckhardt, C.J.; Chance, R.R.; Yee, K.C. Chem. Phys. Lett. 1978, 54, 295.

20. Moran, M.J.; She, C.Y.; Carman, R.L. IEEE J. Quantum Electron. 1975, 11, 259.
21. Shand, M.L.; Chance, R.R.; LePostollec, M.; Schott, M. Phys. Rev. B 1982, 25, 4431.
22. Chance, R.R.; Shand, M.L.; LePostollec, M.; Schott, M. J. Polymer Sci., Polymer Lett. Ed. 1981, 19, 529.
23. Shand, M.L.; Chance, R.R. J. Chem. Phys. 1978, 69, 4482.
24. Shand, M.L.; Chance, R.R.; Silbey, R. Chem. Phys. Lett. 1979, 64, 448.
25. Chance, R.R.; Shand, M.L.; Hogg, C.; Silbey, R. Phys. Rev. B 1980, 22, 3540.
26. Kuzmany, H.; Imhoff, E.A.; Fitchen, D.B.; and Sarhangi, A., Phys. Rev. B 1982, 26, 7109.
27. Mele, E.; in Proceedings of the International Conference on Conducting Polymers,Les Arcs, 1982, J. de Physique (in press).
28. Lauchlan, L.; Chen, S.P.; Etemad, E.; Kletter, M.; Heeger, A.J.; MacDiarmid, A.G. preprint.
29. Levenson, M.D.; Bloembergen N. Phys. Rev. 1974, 10, 4447; Maker, P.D.; Terhune, R.W. Phys. Rev. 1965, 137, A801.
30. Frey, R.; Pradere, F. Opt. Commun. 1974, 12, 98; see also Infrared Phys. 1975, 15, No. 4.
31. Lotem, H.; Lynch, R.T.; Bloembergen, N. Phys. Rev. A 1976, 14, 1748.
32. Sondheimer, F.; Ben-Ephriam, D.A.; Wolorsky, R. J. Amer. Chem. Soc. 1961, 83, 1675.
33. Reimer, B.; Bässler, H. Chem. Phys. Lett. 1978, 55, 315.
34. Meisner, L.B.; Khadzhiiski, N.G. Kvant. Elektron.(Moscow) 1979,6,345.[Sov. J. Quantum Electron. 1979, 9, 199.]
35. Walling, J.C.; Peterson, O.G.; Morris, R.C. IEEE J. Quantum Electron. 1980, 16, 120.
36. Patel, G.M.; Khanna, Y.P. J. Polym. Sci., Polym. Phys. Ed. (in press).
37. Lynch, R.T., Jr.; Kramer, S.D.; Lotem, H.; Bloembergen, N. Opt. Commun. 1976, 16, 372.

RECEIVED June 6, 1983

10

Third-Order Nonlinear Susceptibility in Multilayers of Polydiacetylene

GARY M. CARTER, YUNG JUI CHEN, and SUKANT K. TRIPATHY

GTE Laboratories Inc., Fundamental Research Laboratory, Waltham, MA 02254

The degenerate third order nonlinear susceptibility of a polydiacetylene was investigated through its effect on the index of refraction of the material. The polydiacetylene was deposited to a thickness of 5000 Å a monolayer at a time using The Langmuir Blodgett technique on a silver overcoated grating etched on a silicon wafer. The grating allowed coupling of a freely propagating laser beam into the planar waveguide structure formed by the polydiacetylene on the metal. By changing the intensity of the laser beam a change in coupling angle to the waveguide mode was observed, from which an estimate of the nonlinear index of refraction of the polydiacetylene near the absorption edge was obtained.

The third order nonlinear susceptibility is an important optical property of materials because of its contributions to numerous nonlinear optical processes. (1)(2) With the growing interest in all-optical signal processing it has been proposed (3) recently that χ^3 $(\omega_1, \omega_2, \omega_3)$, and especially the degenerate third order nonlinear susceptibility $\chi^{(3)}$ $(-\omega, \omega, \omega)$ [defined as $\chi^{(3)}(\omega)$], be utilized through its contributions to the changes in dielectric constant ε with optical field strength E:

$$\varepsilon = \varepsilon_1(\omega) + 4\pi\chi^{(3)}(-\omega, \omega, \omega)\, E^2(\omega).$$

$\chi^{(3)}(\omega)$ also contributes to such well known effects as self-focusing, self-trapping and self-bending of light, degenerate four-wave mixing and phase conjugation. (4)(5) A class of polymers, the polydiacetylenes, at wavelengths below the absorption edge have third order nonlinear susceptibilities as large as inorganic semiconductors. (6)(7) Additionally, this class of crystalline polymers provide immense flexibility in structural modification and ease of fabrication. These exciting possibilities and the large

0097–6156/83/0233–0213$06.00/0

value of $\chi^{(3)}(\omega)$ reported for the polydiacetylene poly(bis(p-toluene sulfonate) of 2,4 hexadiyn -1,6 diol) (abbreviated as PTS) has attracted interest for potential all-optical signal processing techniques using this class of materials. (3)

In order to understand the nonlinear susceptibilities of the polydiacetylenes (or any material) one must have a detailed knowledge of the wavefunctions for many levels in the system. Some insight into the problem can be gained by examining $\chi^{(3)}(\omega)$ for a two-level atom or molecule: (8)

$$\chi^{(3)}(\omega) = \frac{N\mu^4}{(E_a - \hbar\omega)^3} \tag{1}$$

Here N = the number of atoms (molecules) per unit volume, μ = transition dipole moment between the two states, E_a = transition energy (separation in energy between the two states), and $\hbar\omega$ is the energy of the incident photon. This formula is valid when the photon energy is near the transition energy. Clearly $\chi^{(3)}(\omega)$ can be large for a large N (e.g., by having a solid), and by having a large μ(as is the case for polydiacetylenes).

For one dimensional systems like the polydiacetylenes, crystalline order is necessary to achieve the maximum value of $\chi^{(3)}$ represented by Equation 1, because randomization of the transition dipole moment direction reduces $\chi^{(3)}(\omega)$ by angle averaging. Additionally, the transition dipole moment itself may diminish as backbone conjugation is decreased due to structural or conformational disorders. Interestingly the method of polymerization in these systems inhibits such disorders and will be discussed in detail in a later section. Although not entirely applicable to the polydiacetylene, Equation 1 also shows that as $\hbar\omega$ approaches E_a, $\chi^{(3)}(\omega)$ is enhanced. As in the case of the linear susceptibility there is no singularity when $E_a = \hbar\omega$ because of the existence of an imaginary component in the resonant denominator. Obviously, the polydiacetylenes are a lot more complicated than the two-level system described above. A more appropriate description is to use a multilevel molecular model, which has been applied with reasonable success in inorganic semiconductors at the off-resonance region. (10) To apply the multilevel model in the resonant region is somewhat tricky because one has to sum all the relevant states with appropriate density of state factors and damping coefficients. We note that a one-dimensional semiconductor model has also been proposed which agrees reasonably well with the experimental data for the nonlinear susceptibilities of the polydiacetylenes at $\hbar\omega \ll E_a$, i.e., for off-resonance. (9)

For this material system we can conclude that the off-resonance optical nonlinearities are not model sensitive. However, as one approaches resonance, the physics of $\chi^{(3)}(\omega)$ is actually quite complicated and not very well known. It is interesting to point out that the first absorption edge in some polydiacetylenes has been attributed to an excitonic state below the band edge.(11) However, no detailed model exists for the contributions to $\chi^{(3)}(\omega)$ by excitons.

Finally, we would like to point out that in the off-resonance region, the response time of the nonlinearity is limited only by the optical pulse width τ, as long as $(E_a - \hbar\omega)/\hbar >> 2\pi(\tau^{-1})$.(8) This is no longer true when collisions (or phonons in solids) are present. For optical frequencies close enough to the absorption edge, the collision induced transitions to the excited state will cause the $\chi^{(3)}$'s response time to be limited by the relaxation time of the excited states. (8)

In this publication, we present the preliminary results of our research effort which is aimed at measuring the magnitude and response time in $\chi^{(3)}(\omega)$ near the fundamental absorption edge in a class of polydiacetylenes that can be grown as controlled thickness crystalline thin films on suitable substrates. The purpose of this research is twofold. First, a detailed investigation of the linear and nonlinear optical properties of the multilayer thin films is expected to yield important correlations between the structural requirements and the properties of interest. Second, this information will be used to guide new synthesis, crystal growth and fabrication techniques.

We first discuss the materials research which includes monomer synthesis, growth of monomer crystalline structures and polymerization in the solid state, yielding the requisite polymer structures. Next, the nonlinear optical experimental research is discussed which includes a novel experimental technique to measure $\chi^{(3)}(\omega)$. Linear and nonlinear optical data obtained for the polydiacetylene films is subsequently presented. Detailed theoretical analysis relating the data to $\chi^{(3)}(\omega)$ and subsequently to its molecular basis will be discussed in a later publication.

Choice of the Monomer and Monomer Synthesis

The diacetylene monomer employed in the thin film growth technique pioneered by Langmuir and Blodgett (12) must have a strongly polar "head group" and a nonpolar "tail." The monomer we have used in our studies, $CH_3 - (CH_2)_{15} - C \equiv C - C \equiv C - (CH_2)_8 - COOH$, has a long alkyl group as the nonpolar "tail." This class of monomers has been extensively studied for the structural aspects of their molecular assemblies. (13) Apart from the acid group as the polar end, other end groups such as alcohols and various alkali metal salts of the acids have also been investigated.

The monomer used in our studies was synthesized by Cadiot-Chodkiewicz Coupling (14) of 1-iodooctadecyne and 10-Undecynoic acid. 10 Undecynoic acid was used as received from Farchan Labs and the iodooctadecyne was prepared from 1-octadecyne by the procedure provided below. Two grams of 1-octadecyne (0.008 mole) in 15 ml of hexane was added by drops to 5 ml of a 1.8 molar n-butyl lithium/hexane solution. The bright yellow solution was refluxed for 40 minutes and allowed to cool. A slight excess of iodine (0.009 mole) in 90 ml of hexane was then added by drops to the solution, which was then refluxed for one hour. The resulting purple solution was allowed to stir overnight. The solution was subsequently acidified with glacial acetic acid to a pH of 4. The hexane layer was washed with water and a 10 percent aqueous solution of sodium thiosulfate and dried over anhydrous magnesium sulfate. The product was obtained by evaporating the solvent at reduced pressure. The IR spectrum of the product showed no evidence of unreacted 1-octadecyne.

For the coupling reaction, all manipulations were done in the dark and under inert atmosphere. 0.014 gram (0.00014 mole) of cuprous chloride was dissolved in 2 ml of ethyl amine (70%) in water and added to 0.76 gram (0.0045 moles) of 10-undecynoic acid in 35 ml of methanol. Enough hydroxylamine hydrochloride was then added to the flask to turn the solution color from bright blue to clear. 1.70 grams (0.0045 moles) of 1-iodooctadecyne was dissolved in 40 ml of tetrahydrofuran and added by drops to the flask while rapidly stirring the contents. As the addition proceeded, the solution turned yellow and then green at which time the catalytic copper species was regenerated by adding hydroxylamine hydrochloride. This step was repeated as often as necessary to maintain the yellow color of the solution. It is important to note that the reaction stops if the pH drops below 7.

After the completion of the reaction, the solution was acidified to a pH of 4 with 2N sulfuric acid, followed by the addition of 50 ml of water. The solution was then extracted several times with ether. The extractant was dried over magnesium sulfate and the solvent was removed by evaporation at reduced pressure. Cold petroleum ether was then added to the resultant oily material to precipitate the product. The product was further washed with cold (10° C) petroleum ether and recrystallized several times from warm petroleum ether. The melting point of the final product was 73.5^{o} C. Infrared spectrum of the product showed major absorption peaks relevant to the pure monomer (Figure 1). Under UV radiation, white flakes of the monomer solid turned deep blue (partial polymerization).

Thin Film Growth and Multilayer Assemblies

The diacetylene monomers can be spread as a monomolecular layer on the air-water interphase of a Langmuir-Blodgett trough. The molecules are well oriented because of the preferences of the hydrophilic head group of the monomer to be in coordination with the subphase. The monomers can be spread using a good solvent like chloroform.

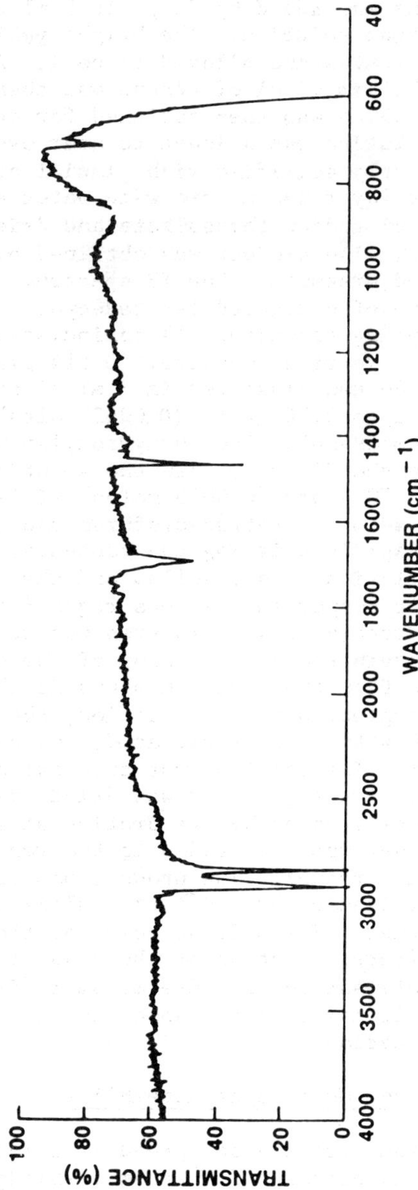

Figure 1. IR spectrum of the diacetylene monomer on sodium chloride plates. All the peaks are properly assigned and indicate high purity of the monomer.

The Langmuir-Blodgett trough essentially consists of a teflon coated rectangular trough. A fixed barrier with an attached transducer at one end contains the monomer molecules within a surface region and also records the surface pressure. A barrier on the other end of the trough that can be moved through the length of the trough is used to reduce the surface area available per monomer and thus get them into an organized continuous monolayer film. A schematic of a LB apparatus is shown in Figure 2. The pressure-area isotherms give an indication of the extent of packing and the "quality" of the film.

The pressure-area isotherm of the diacetylene monomers is shown in Figure 3. The arrow shows the point at which the film has buckled. The monomer molecules are optimally packed at a surface pressure of ≃30 dynes/cm. At this stage polymerization in the monolayer can be carried out by UV radiation. In our laboratory a specially designed large area UV lamp (252 nm) covering the whole monolayer film is employed for this purpose. There are several factors that effect the "quality" of the monolayer films. The monomer solution should be free of impurities and the trough should be clean and filled with distilled deionized water. The manner of deposition of the monomer solution onto the trough is crucial. In our experiments the monomer solution was deposited near the fixed barrier while the movable barrier was being moved away. This has an orienting effect on the monomer and limits any agglomeration. Also, the rotational freedom of the monomer molecule is maintained by saturating the region above the interphase with the solvent vapour as the monomers get into a close packed arrangement.

As the film is polymerized there is some reduction (≃10%) in the total surface area of the films. The film can be transferred to a suitable substrate by dipping the substrate through the film and subsequently pulling it off while a fixed film surface pressure is maintained. Thus, in each "dipping" two monolayers are transferred onto the substrate. By successive dippings a multilayer assembly can be fabricated on the substrate. The preferred substrate for the multilayer dipping should have a hydrophobic character. This leads to an excellent adhesion between the aliphatic tails of the monomer to the substrate.

Multilayer assemblies of ≃0.5 μm thickness were built up on quartz plates and the films were characterized in detail using spectroscopic (vis. IR) and electron microscopic techniques. Figure 4 is an optical absorption spectrum from a multilayer assembly and shows the sharp absorption in the visible characteristic of the polydiacetylenes. Electron diffraction reveals a crystalline layered structure. However, registry between layers is less than perfect. Electron diffraction from a few layers indicates a strong possibility for growing well-oriented structures, and this is being pursued in our laboratory.

For the elucidation of the optical and nonlinear optical properties of the polydiacetylene layers, we have employed a wave-

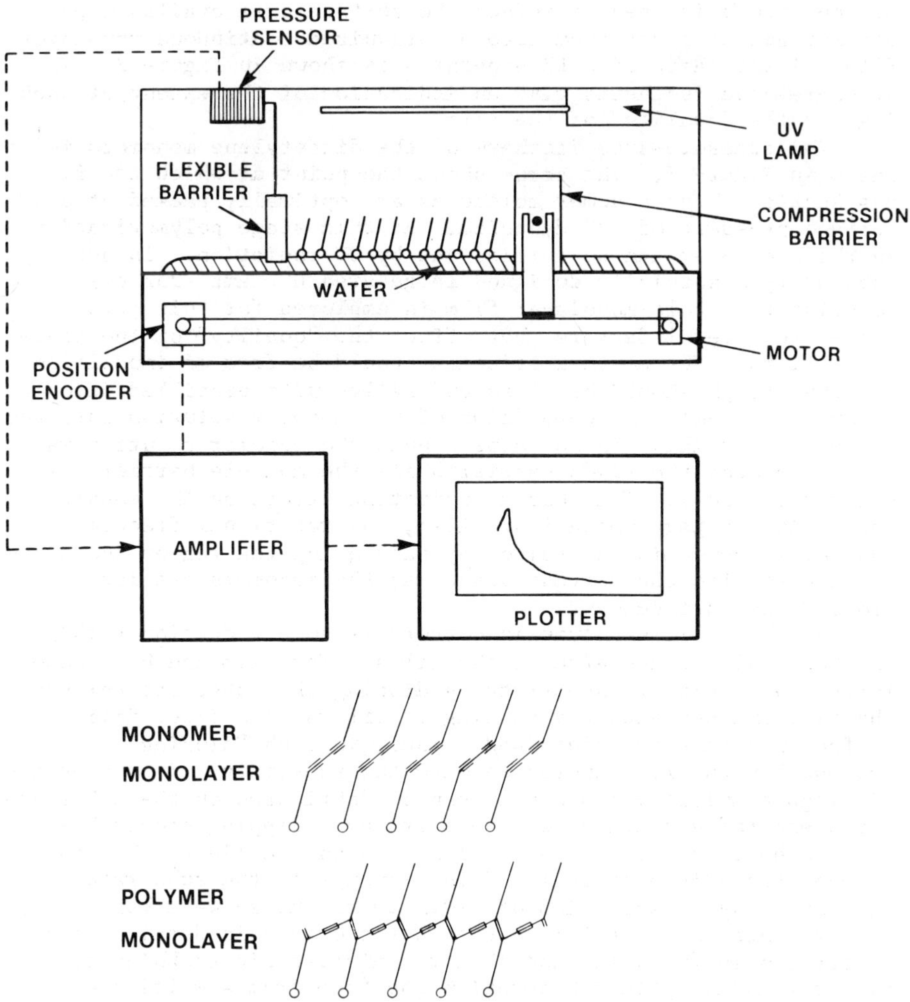

Figure 2. Schematic of the Langmuir Blodgett film balance and monolayer deposition at the air–water interphase.

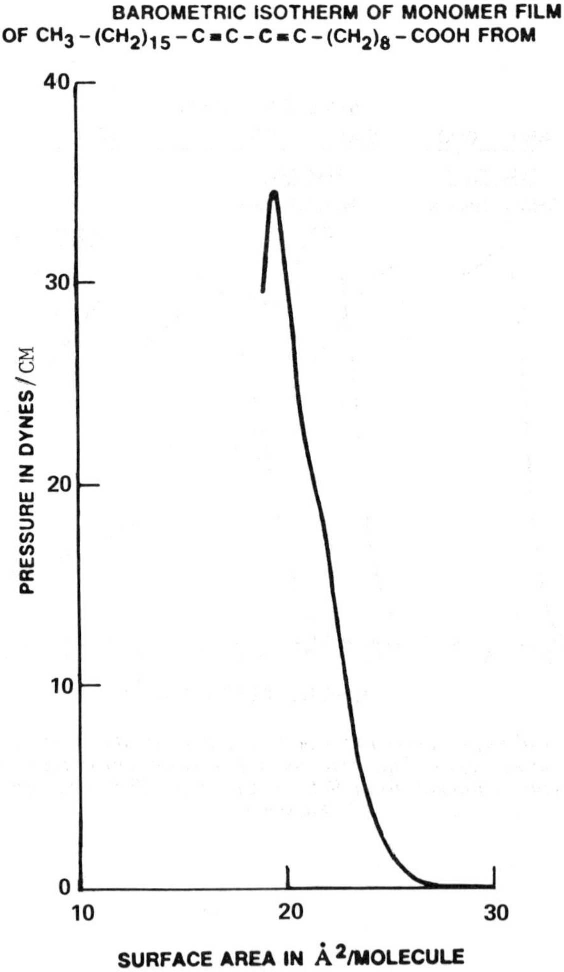

Figure 3. Pressure area isotherm at room temperature for the diacetylene under investigation.

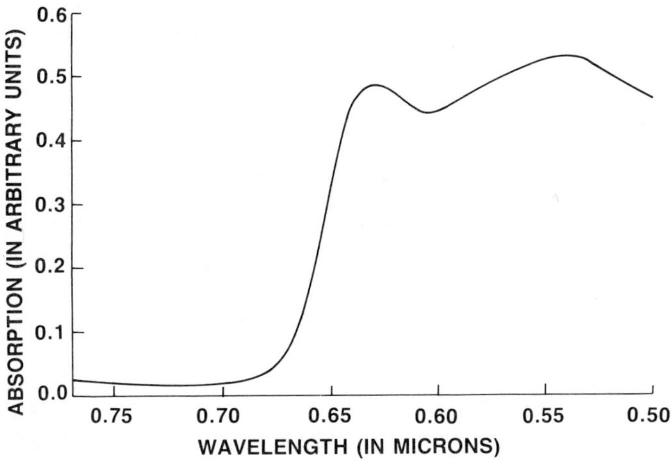

Figure 4. Absorption spectrum in the visible region of the multilayer polydiacetylene film deposited on a quartz substrate. Note the sharp absorption edge.

guide coupling technique which will be discussed in the subsequent section. For this purpose multilayer assemblies were fabricated on a silver coated grating etched on a silicon substrate. The grating was fabricated using silicon planar technology and a silver layer was subsequently overlaid by vapour deposition.

Measurement Technique

The experimental technique used to measure the third order nonlinear susceptibilities of polydiacetylenes was developed recently at our laboratory. (15) This technique utilizes the intensity-dependent dispersion relation of confined modes of PDA films and determines $\chi^{(3)}$ by measuring the change of coupling angles with input intensity. Since this nonlinear coupling technique determines the change of index of refraction of the system (i.e., PDA film) with input light intensity, it yields information about both the magnitude and the sign (with respect to $\chi^{(1)}$) of $\chi^{(3)}$. In the cases that the input wavelength is near the absorption edge of PDA, where many nonlinear mechanisms could be contributing to the third order nonlinearity, this knowledge of both the sign and the magnitude of $\chi^{(3)}$ is particularly valuable for identifying the dominating nonlinear mechanisms. The nonlinear coupling technique can also be used with the pump and probe technique to obtain further information about the relaxation time of various nonlinear mechanisms.

Since the confined modes (whether they are waveguide modes or surface plasmons) are nonradiative (i.e., their wavevector parallel to the interface, k_{11}, is greater than the wavevector of the photon in air), one must increase k_{11} of the input photon to couple into the confined mode. This can be accomplished by using either a prism or a grating coupling technique. (16)(17) For reasons of fabrication ease, coupling stability, lower background noise and ease of alignment, we chose the grating coupling method for our experiments.

The experimental setup is shown in Figure 5. Our light source is derived from a high power YAG pumped pulsed dye laser (100 mj/pulse, 10 pulse/sec and a pulse width ≃5 nsec). The dye laser is tunable throughout the visible spectrum. When a high pressure H_2 cell is used to Raman-shift the wavelength (through simulated stokes Raman processes), the dye laser can be tuned beyond 1.3 μm. An x-ray spectrogoniometer (having an angular resolution of 0.01°) was used for the reflection measurements. The holographic gratings (periodicity ≃4854 angstroms) were etched on the Si substrates using a dry-etch technique. (18) A micron thick layer of silver was subsequently evaporated on to the grating in a vacuum chamber at 10^{-6} torr. As discussed earlier, the PDA layers were deposited onto the silvered grating using a dipping apparatus subsequent to polymerization on the air-water interface.

As shown in the inset of Figure 5, the polarized laser beam was incident on the PDA surface at an incident angle θ. The grating grooves are set vertically (normal to the sagittal plane). Both vertically polarized (TE) and horizontally polarized (TM) incident light were studied. In the TE case, two (one at wavelengths > 8500 angstroms) waveguide modes in the PDA films were observed. In the TM case, a surface plasmon (at the Ag-PDA interface) and two or one TM waveguide mode were observed. We will limit our discussion only to the TE waveguide modes. The detail discussion of TM mode results will be reported in a later publication.

Since the PDA films were deposited using the Langmuir-Blodgett technique, the polymer chains in each layer were oriented along the plane. Thus the PDA film can be treated as an isotropic medium for the TE modes. Figure 6 shows a typical angular scan of reflected intensity result at 7500 angstroms. We note that the coupling efficiency and the attenuation constant of the $n = 1$ mode is approximately half of that of the $n = 2$ mode (from the width and size of the reflectivity dip). Since there are two possible grating coupling regions for the waveguide mode: the PDA film-Ag interface and the air-PDA film interface, the result suggests that the dominate coupling region is at the air-PDA film interface (i.e., the PDA film surface kept the grating profile). Figure 7 shows the field distributions for the $n = 1$ and $n = 2$ modes using reasonable optical parameters (index 1.51, d=5500 angstroms) for the PDA film. We note that the field amplitudes at the air-PDA film interface for both modes are much larger than the corresponding field amplitude at the PDA film-Ag interface, which confirms our previous argument that the PDA-air interface is a better coupling region if a comparable grating exists.

For the nonlinear experiments, one changes the input laser intensity and, hence, changes the dielectric constant of the film via the effect of $\chi^{(3)}(\omega)$. This change in dielectric constant causes a change in coupling angle (i.e., shifts the position of the reflectivity minimum) by $\Delta\theta$. By using the linear coupling coefficients and knowledge of the mode profile, one can deduce $\chi^{(3)}(\omega)$ from $\Delta\theta$. (15) We present preliminary data (Figure 8) at $\lambda = 6700$ angstroms in the absorption region of the PDA film. For a $E_L = .7\text{x}10^{-6}$j and $E_H = 7\text{x}10^{-6}$j (E_L, E_H are the incident laser energies for the two curves), the measured $\Delta\theta = -1.7^o$. We note the angular shift is negative, which indicates that $\chi^{(3)}$ is negative with respect to the zero field index.

Although a great deal of data analysis is needed to obtain a value of the optical nonlinearity from this measurement, we can estimate the order of magnitude of the value. The estimate comes from a comparison of this data with that taken on a similar experiment using Si as the nonlinear material. (15) The two experiments used approximately the same laser power and beam geometry, and the linear reflectivity curves were similar in shape and size (the minimum value of reflectivity). Neglecting differences

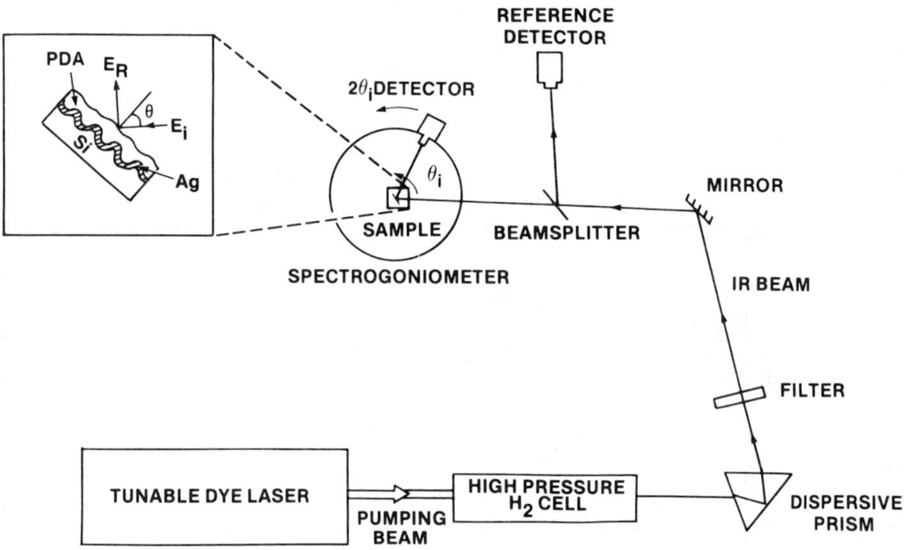

Figure 5. Experimental setup for the intensity-dependent nonlinear coupling measurement. The sample, as shown in the inset, is mounted on an x-ray spectrogoniometer and the polarized light is incident on the PDA film surface at an angle, θ.

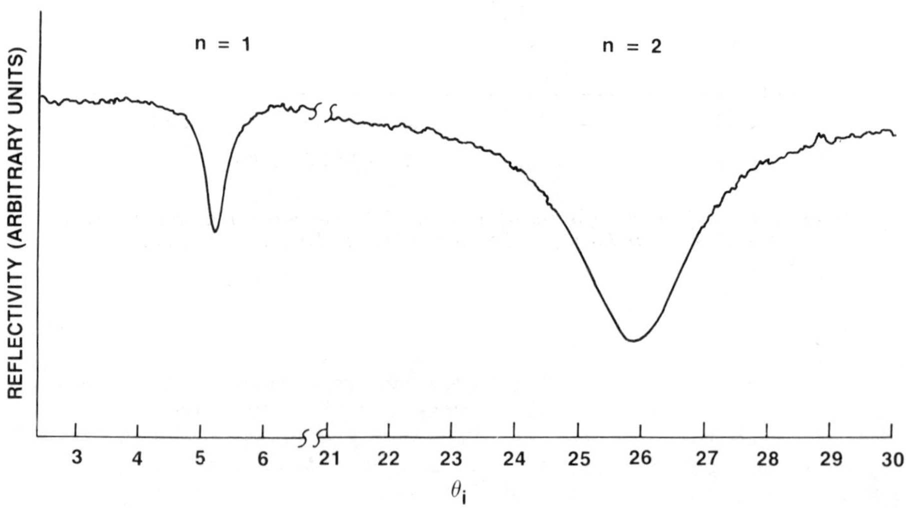

Figure 6. Reflectivity vs. incident angle θ at wavelength 7500 Å for TE polarized light. Two TE modes (n = 1 and 2) were observed, as designated.

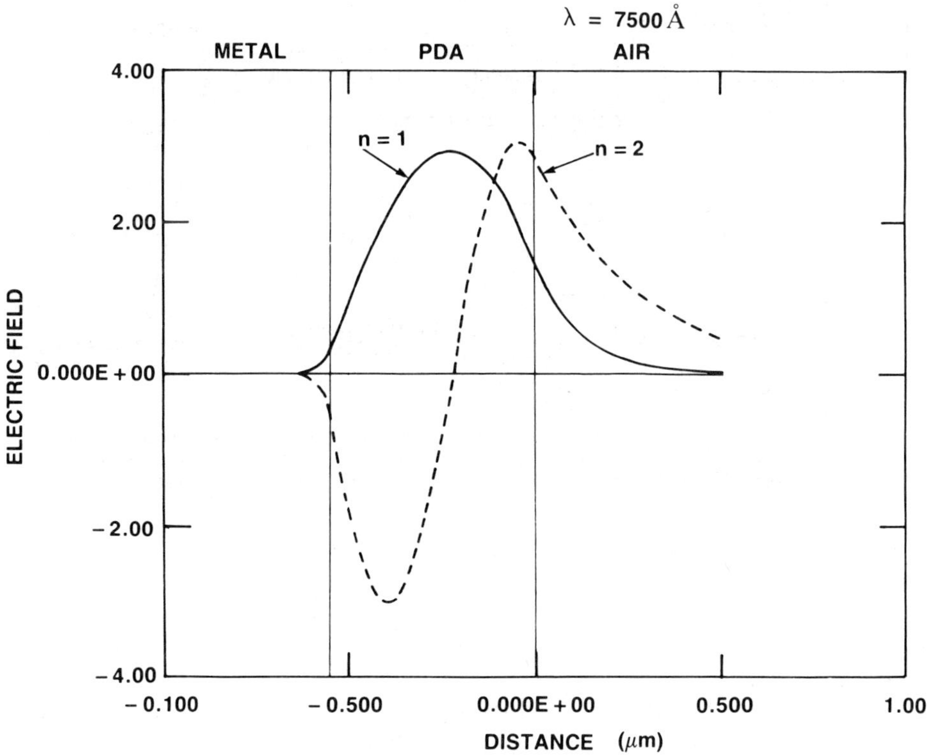

Figure 7. Field distribution for the n = *1 and* n = *2 TE waveguide mode for an air–PDA film–silver system for* n_{PDA} = *1.51,* d = *5500 Å and a Drude-model metal.*

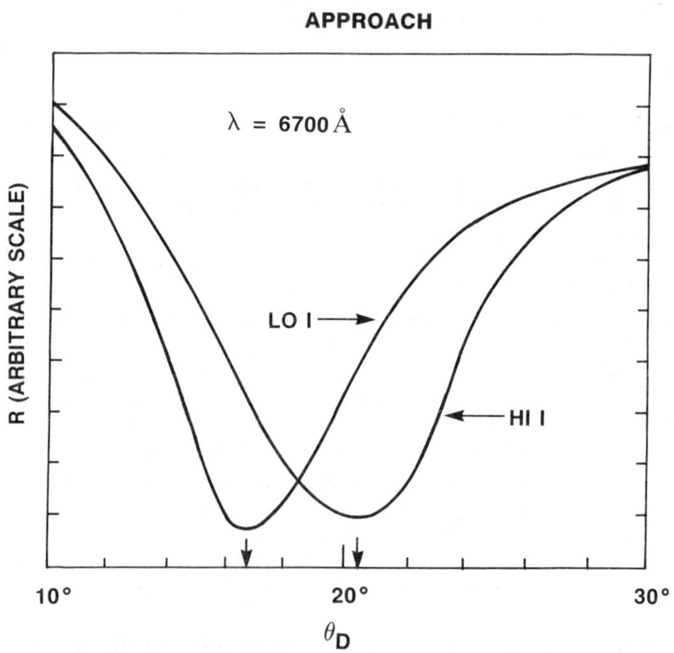

Figure 8. Nonlinear coupling result (reflectivity vs. angle) for PDA film at λ = 6700 Å. The high and low intensity (Hi I and Lo I) measurements were done under the same experimental condition and $E_L = 0.7 \times 10^{-6} j$ *and* $E_H = 7 \times 10^{-6} j$ *(details are discussed in the text). Note* θ_D *is the detector angle =* $2\theta_i$*.*

in mode shape between the two cases, the observed shift was approximately the same in each case, thus indicating the value for n_2 for this PDA at 6700 angstroms $\approx 10^{-5}$ $(MW/cm^2)^{-1}$, where the $\approx$ sign means "of the order." The Si data was also taken in the absorption edge of that material, and the nonlinearity observed by this technique agreed well with that obtained by other techniques. (15)(19)

We are currently extending our study to other wavelengths, especially in the transparent region of the film, $\lambda > 7000$ angstroms. By examining the linear coupling coefficients and mode shape as a function of wavelength, we should be able, once $\Delta\theta$ vs. λ is measured, to determine the $\chi^{(3)}$ vs. λ.

Conclusion

Linear and nonlinear optical properties of a class of polydiacetylenes that can be grown as large area and controlled thickness thin films have been investigated. This involved an integrated research effort including monomer synthesis, thin film growth and new measurement techniques.

The ease of grating coupling to the polydiacetylene waveguide structure and our observation of the optical nonlinearity in the absorptive region of the material, along with the sensitivity of the measurement technique, has opened up the possibility of a detailed study of n_2 versus λ in the absorptive and transparent regions, as well as in the time domain. From the narrow angular width of the reflectivity dips for these planar waveguides, observed electron microscopy and diffraction data and the observable optical nonlinearity, we conclude that there is a high degree of order in thin films deposited to form the waveguide.

The success of this research is guiding our materials efforts into increasing chain packing density and orientation to increase n_2, including a variety of new thin film deposition techniques.

One should note that to realize practical devices based on these materials, in addition to the research described in the preceding, research in processing and fabrication techniques related to the PDA's is necessary. In fact, this is also true for inorganic semiconductor materials.

Acknowledgments

The authors would like to thank Dr. Nick Economou of Lincoln Laboratory for fabricating the Silicon gratings, and Jacque Georger, Joe Gormley and Mike Rooney for expert technical assistance. Guidance in monomer synthesis and many fruitful discussions with Michael Rubner and Daniel Sandman of our laboratory are gratefully acknowledged.

Literature Cited

1. Frohlich, D. "Advances in Solid State Physics: Festkoer Probleme"; Pergamon: New York, 1981; Vol. XXI, p. 363.
2. Levenson, M. D.; Song, J. J. "Coherent Nonlinear Optics," Feld, M. S. and Letokhov, V. S., eds.; Springer: New York, 1980; Chapter 7.
3. Smith, P. W. "On the Physical Limits of Digital Optical Switching and Logic Elements," Bell System Techn. J. 1975 (1982), 6.
4. Shen, Y. R. Rev. Mod. Phys. 1976, 48, 1.
5. Pepper, D. M. Opt. Eng. 1982, 21, 156.
6. Sauteret, C; Hermann, J-P; Frey, R.; Pradere, F.; Ducuing, J.; Baughman, R. H.; Chance, R. R. "Optical Nonlinearities in One Dimensional-Conjugated Polymer Crystals," Phys. Rev. Lett. 1976, 36, 956.
7. Hermann, J.P.; Smith, P. W.; "Nonlinear Fabry-Perot Containing the Polydiacetylene PTS," Digest of Technical Papers, Eleventh International Quantum Electronics Conference, Boston, MA, June 23 - 26, 1980, published by IEEE, p. 656.
8. Hanna, D. C.; Yuratich, M. A.; Cotter, D. "Nonlinear Optics of Free Atoms and Molecules"; MacAdam, D. L., ed.; Springer-Verlag: Berlin, 1979; Section 2.6.2.
9. Agrawal, G. P.; Cojan, C.; Flytzanis, C.; "Nonlinear Optical Properties of One-Dimensional Semiconductors and Conjugated Polymers," Phys. Rev. 1978, B17, 776.
10. Jha, S. S.; Bloembergen, N.; Phys. Rev. 1968, 171, 891.
11. See, for example, Philpott, M. R. "Optical Reflection Spectroscopy of Organic Solids," Ann Rev. Phys. Chem. 1980, 31, 97.
12. Blodgett, K. B.; Langmuir, I. Phys. Rev. 1937, 51, 964.
13. Day, D. R., Ph.D. Thesis, Case Western Reserve University, 1980.
14. Chodkiewicz, W. Ann. Chim. 1957, 2, 819.
Egleton, G.; McRae, W. "The Coupling of Acetylenic Compounds," Advances in Organic Chemistry 1963, 4, 225.
15. Chen, Y. J.; Carter, G. M. "Measurement of Third Order Nonlinear Susceptibilities by Surface Plasmons," Appl. Phys. Lett. 1982, 41, 307.
16. Tamir, T. in "Integrated Optics"; Tamir, T., ed.; Springer: New York, 1978; Chap. 3.
17. Burstein, E.; Chen, W. P.; Chen, Y. J.; Harstein, A. J. Vac. Sci. Technol. 1974, 11, 1104.
18. Geis, M. W.; Lincoln, G. A.; Efremow, W.; Piacentini, W.J. "A Novel Anisotropic Dry Etching Technique," J. Vac. Sci. Technol. 1981, 19, 1390.
19. Jain, R. K.; Klein, M. B. Appl. Phys. Lett. 1979, 35, 454.

RECEIVED June 23, 1983

11

Epitaxial Growth of Polydiacetylenes

SCOTT E. RICKERT, JEROME B. LANDO, and STEPHEN CHING

Case Western Reserve University, Department of Macromolecular Science, Cleveland, OH 44106

The two-step process of epitaxial polymerization has been applied to symmetrically substituted diacetylenes. First, the monomers have been crystallized epitaxially on alkali halides substrates from solution and the vapor phase. The oriented monomer crystals are then polymerized under the substrate's influence by gamma-irradiation. The diacetylenes in this study are 2,4-hexadiyn-1,6-diol (HD) and the bis-phenylurethane of 5,7-dodecadiyn-1,12-diol (TCDU). The polydiacetylene crystal structures and morphologies have been examined with the electron microscope. Reactivity and polymorphism are found to be controlled by the substrate.

Epitaxial polymerization is a general process applicable to monomers that may be polymerized in the solid state. The study of disulfurnitride vapor phase crystallization on alkali halides with thermal polymerization to polythiazl, $(SN)_x$, has shown that substrate controlled reaction led to three new crystal phases of $(SN)_x$ (1-2). Diacetylenes are monomers that can be polymerized from the monomer crystals to varying degrees of conversion and crystallinity depending on the nature of the substituents and their packing within the monomer lattice(3). Furthermore, the polymer backbone may adopt either an acetylenic $\leftarrow$RC-C≡C-CR$\rightarrow$ or a butatriene $\leftarrow$RC=C=C=CR$\rightarrow$ bonding.

The urethane-substituted polydiacetylenes exhibit thermochromic transition with low and high temperature crystal phases favoring acetylenic and butatriene backbone, respectively (4-6). Our interest in the application of epitaxial polymerization to diacetylenes has been the possibility of substrate control over orientation, structure, and the single crystal nature of thin films.

0097-6156/83/0233-0229$06.00/0

Experimental

2,4-hexadiyne-1,6-diol (DMDA) from Farchan Laboratories was recrystallized from toluene. Solutions up to 1 wt. % in toluene were prepared for epitaxial crystallization. Bis-phenylurethane of 5,7-dodecadiyne-1, 12-diol (TCDU) was supplied by Allied Chemical Co. The monomer was extracted from the partially polymerized lattice with acetone. The monomer was dissolved in ethyl acetate to make a 0.4 wt. % solution.

Table I. TCDU

	Phase 1	Phase 2
Monomer	a = 0.708 nm	a = 1.160 nm
	b = 3.397 nm	b = 1.897 nm
	c = 0.523 nm	c = 0.519 nm
	β = 115.85°	γ = 91.0°
Polymer	a = 0.623 nm	a = 1.184 nm
	b = 3.903 nm	b = 1.987 nm
	c = 0.491 nm	c = 0.495 nm
	β = 106.85°	γ = 94.94°

Epitaxial crystallization was accomplished by the immersion of a preheated substrate (the substrates for epitaxial crystallization were single crystals of alkali halides from Harshaw Chemical Co.) into the monomer solution followed by its in-situ cleavage to expose two fresh (100) surfaces.

Table II. Diol Unit Cell Parameters

	a	b	c	β
Monomer	.409	1.60	.477	106.3
Polymer	.411	1.61	.482	106.1
Polymer/KBr	.421±.04	1.02±.10		

units in nm

The isothermally crystallized monomers were polymerized with 2·4 Mrad γ-irradiation. The polymer film was removed from the substrate after C , Pt coating by dissolving the salt. A JEOL JEM 100B electron microscope was used to examine the samples.

Results and Discussion

Large domains of oriented single crystals of poly(TCDU) and poly(DMDA) were produced on the alkali halide surface. Figures 1 and 2 show the typical elongated platelet morphology. Selected

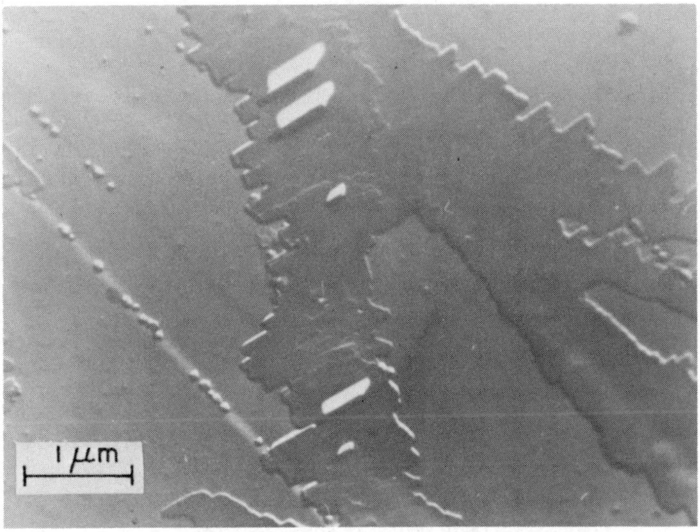

Figure 1. Poly(TCDU) polymerized on KCl and deposited from a 0.4 wt% ethyl acetate solution at 30 °C.

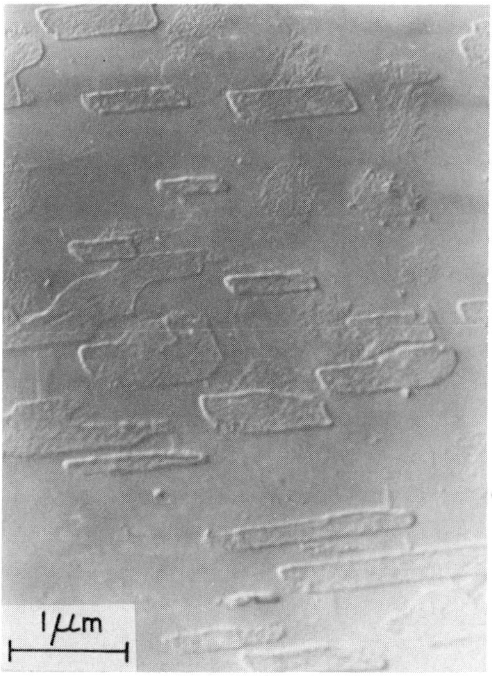

Figure 2. DMDA polymerized on KBr and deposited from a wt% toluene solution at 78 °C.

area electron diffraction from such domains gave single crystalline pattern as in Figures 3 and 4. The unit cell parameters are given in Tables 1 and 2.

Poly(TCDU) in thin film exists in the acetylenic phase 2, as opposed to the butatriene phase 1 found in bulk crystals polymerized from macroscopic monomer crystals (7). Polymerization to completion in thin film, thereby removing the constraint of the monomer lattice, could account for the acetylenic phase.

The orthogonal projection of the epitaxial poly(DMDA) could not be indexed using the unit cell data for the bulk polymerized crystal (8). However, poly(DMDA) cannot usually be polymerized to completion or to high crystallinity in the bulk due to cross-linking. The use of an epitaxial substrate may have controlled the polymerization process that led to oriented single crystals.

Conclusion

Two diacetylenes have been epitaxially polymerized as thin films in contact with alkali halide substrates. These films in contact with alkali halide substrates. These films consisted of highly oriented single crystals aligned along both <110> directions of the substrate. The structures of both poly(TCDU) and poly(DMDA) were modified by this technique and, in all cases, highly crystalline near-perfect films were achieved.

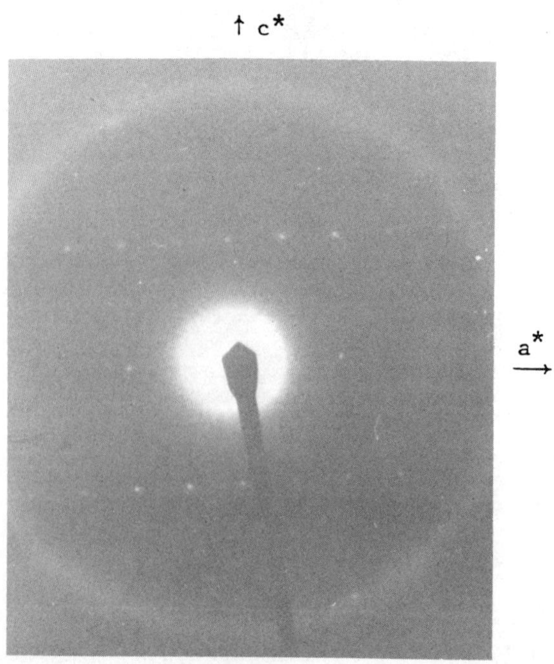

Figure 3. Electron diffraction of poly(TCDU) formed under conditions specified in Figure 1.

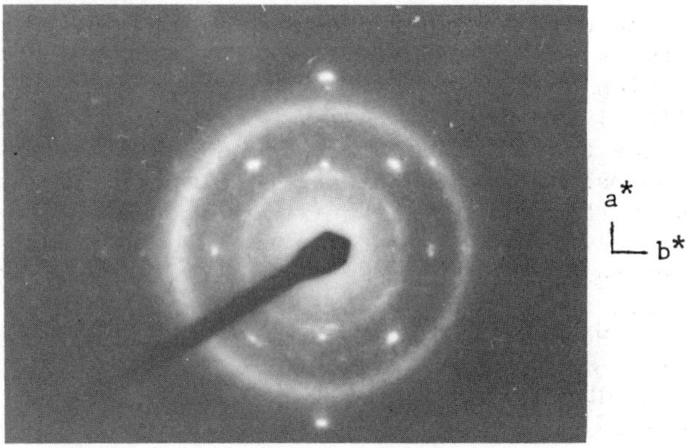

Figure 4. Electron diffraction of poly(DMDA) from Figure 2.

Acknowledgments

The authors would like to acknowledge the support of Army Research Office Grant No. DAAG 29-80C-0099 for financial assistance of this work.

Literature Cited

1. Rickert, S. E; Lando, J. B.; Hopfinger, A. J.; Baer, E. Macromolecules 1979, 12, 1053.
2. Rickert, S. E.; Ishida, H.; Lando, J. B.; Koenig, J. L.; Baer, E. J. Appl. Phys. 1980, 10, 5194.
3. Baughman, R. H. J. Polym. Sci., Polym. Phys. Ed. 1974, 12, 1511.
4. Chance, R. R.; Baughman, R. H.; Miller, H.; Eckhardt, C. J. J. Chem. Phys. 1977, 67, 3616.
5. Miller, H.; Eckhardt, C.J. Mol. Cryst. Liq. Cryst. 1978, 45, 313.
6. Eckhardt, H.; Eckhardt, C. J. J. Chem. Phys. 1979, 70, 5498.
7. Enkelmann, V.; Lando, J. B. Acta Cryst. 1978, B34, 2352.
8. Baughman, R. H. J. Appl. Phys. 1972, 43, 4362.

RECEIVED June 6, 1983

12

Nonlinear Electro-optic and Dielectric Properties of Flexible Polymers

GARO KHANARIAN[1] and ALAN E. TONELLI

Bell Laboratories, Murray Hill, NJ 07974

Experimental and theoretical results are presented for four nonlinear electrooptic and dielectric effects, as they pertain to flexible polymers. They are the Kerr effect, electric field induced light scattering, dielectric saturation and electric field induced second harmonic generation. We show the relationship between the dipole moment, polarizability, hyperpolarizability, the conformation of the polymer and these electrooptic and dielectric effects. We find that these effects are very sensitive to the details of polymer structure such as the rotational isomeric states, tacticity, and in the case of a copolymer, the comonomer composition.

The nonlinear optical and dielectric properties of polymers find increasing use in devices, such as cladding and coatings for optical fibres, piezoelectric and optical fibre sensors, frequency doublers, and thin films for integrated optics applications. It is therefore important to understand the dielectric, optical and mechanical response of polymeric materials to optimize their usage. The parameters that are important to evaluate these properties of polymers are their dipole moment $\underset{\sim}{\mu}$, polarizability $\underset{\sim}{\alpha}$, hyperpolarizabilities $\underset{\sim}{\beta}$ and $\underset{\sim}{\gamma}$, and their relationship to the conformation of the polymer. To elucidate these parameters we have undertaken experimental and theoretical studies of a number of nonlinear dielectric and electrooptic effects of single polymer chains.

The electrical and optical properties of polymers such as the dipole moment and hyperpolarizabilities are tensors of rank 1,2,3,4 etc. In order to extract as much information as possible about the elements of these tensors, it is necessary to orient the polymers; in this paper the orientation is achieved by application of a strong electric field. The degree of orientation is probed by a weak analyzing electric field, this field being either static or of very high frequency from a light source. The combination of strong orienting and weak analyzing fields gives rise to a number of electrooptic and dielectric experiments. In this paper we shall discuss the Kerr effect (1–5) dielectric polarization and saturation (6,7) electric field induced second harmonic generation (8) and electric field induced light scattering. (9)

[1]Current address: University of Massachusetts, Polymer Research Institute, Amherst, MA 01003.

0097–6156/83/0233–0235$06.00/0

The above effects have been studied in the gas and solution phase for rigid organic molecules and their electrical properties have been readily deduced. In the case of flexible molecules and polymers there is an additional complication in the interpretation of the experimental results, due to the existence of many conformations. Flory (10) has shown how we may calculate the average of any configurationally dependent property of a polymer. The method rests upon two important assumptions. Firstly, it is assumed that each rotatable bond has a discrete number of states, usually three corresponding to the trans, gauche plus and minus conformations, respectively. This is called the rotational isomeric state approximation (RIS). Furthermore it is assumed that only nearest neighbor interactions are important in determining the probabilities of the bond rotational states. Secondly it is assumed that the electrical property to be calculated, say $\underset{\sim}{\alpha}$, is the tensorial sum of noninteracting contributions $\underset{\sim}{\alpha}_i$ from groups or segments that constitute the polymer. The bond additivity approximation (BAA) appears to work for polymers dissolved in isotropically polarizable nonpolar solvents. However in the gas phase, BAA has been shown to be incorrect by Ward and coworkers (11). It has been speculated that the solvent provides a symmetrical environment in which local electric fields at a given bond caused by adjoining bonds, are cancelled by fields due to solvent molecules. Thus assuming the correctness of the RIS and BAA models, the configurational average $<g(\tau)>$ over all internal degrees of freedom τ is given by

$$<g(\tau)> = Z^{-1}J^{*}G_{1}^{n}J \qquad (1)$$

where Z is the partition function, G_1^n denotes a serial product of generator matrices for each bond in the chain (1 to n), and J^* and J are row and column vectors. The generator matrix G_i contains information about the geometry, rotational states and their energies and electrical parameters of the ith bond of the polymer. Flory's book[10] gives the generator matrices for the configurational averages occuring in this paper.

Kerr Effect

The Kerr effect is the birefringence induced in a medium by an external electric field (12). From such an experiment we deduce the molar Kerr constant mK, thus

$$mK = \frac{2\pi N_A}{135}\left(\frac{<tr(\hat{\underset{\sim}{\alpha}}^{o}\hat{\underset{\sim}{\alpha}})>}{kT} + \frac{<\underset{\sim}{\mu}^{T}\hat{\underset{\sim}{\alpha}}\underset{\sim}{\mu}>}{k^2T^2}\right)$$

$$\hat{\underset{\sim}{\alpha}} = \underset{\sim}{\alpha} - \bar{\alpha}\underset{\sim}{I} \qquad (2)$$

In eqn (2) N_A, k and T are Avogadro's number, Boltzmann's constant and temperature respectively. $\hat{\underset{\sim}{\alpha}}$ and $\hat{\underset{\sim}{\alpha}}^o$ are the optical and static anisotropic polarizability tensors and $\bar{\alpha}$ and $\underset{\sim}{I}$ denote the average polarizability and identity matrix, $\underset{\sim}{\mu}$ and $\underset{\sim}{\mu}^T$ are the dipole moment and its transpose, respectively. In eqn (2) the first and second terms are the induced and permanent dipole contributions to the orientation.

We built a sensitive Kerr effect apparatus (1) which used a phase sensitive technique to detect birefringences of the order of 10^{-5} to 10^{-6} radians. Polymers were dissolved in nonpolar solvents in the concentration range 0.5 to 2% w.w. Using procedures for extrapolation to infinite dilution, we obtained experimental mK values. The polymers that we studied were the α,ω dibromoalkanes (1), poly(oxyethylene glycol) (2) and its oligomers, poly(vinylchloride) (4) and its oligomers, the copolymer poly(styrene-co-p-bromostyrene) (3) and fluorinated polymers (5) including poly(vinylidene fluoride). We also calculated mK from eqn (1) and (2), using the RIS model applicable to the polymer under study.

Our conclusions are that a) theta conditions are not necessary to obtain experimental mK values unperturbed by excluded volume interactions, b) the RIS and BAA approximations are applicable and c) mK is a very sensitive to the details of the RIS model, the tacticity and the composition of a copolymer.

In table I we present the molar Kerr constants and mean square dipole moments $<\mu^2>$ of three fluorinated polymers, poly(trifluoroethylene) (PF_3E), poly(vinylidene fluoride) (PVF_2) and poly(fluoromethylene) (PFM), dissolved in p-dioxane. The results show the sensitivity of mK to the degree and type of fluorination varying over an order of magnitude and also changing sign. Calculations of mK and $<\mu^2>$ for comparison are in progress (5).

Table I

$mK/x(x10^{-12}\ cm^7\ SC^{-2}mol^{-1})$ and $<\mu^2>/x(x10^{-36}\ SC^2cm^2)$ per monomer for poly(trifluoroethylene) PF_3E, poly(vinylidene fluoride) PVF_2 and poly(fluoromethylene) PFM, measured in p-dioxane solution at 25°C.

Polymer	Structure	mK/x	$<\mu^2>/x$
PF_3E	$(-CHF-CF_2-)_x$	-9.1	0.86
PFM	$(-CHF-)_x$	~0	0.31
$PVF_2^{a)}$	$(--CF_2-CH_2-)_x$	14	2.38

a) Measured at 60°C

In figure 1 we present the experimental and calculated mK values of the copolymer poly(styrene-co-p-bromostyrene). From this study (3) we were able to show unequivocally that the tacticity of this polystyrene sample is $p_r = 0.55$, where p_r is the probability of racemic dyad replication.

Electric Field Induced Light Scattering

Light scattering from macromolecules is used routinely to obtain molecular weights, radii of gyration and polymer-solvent and polymer-polymer interaction parameters. A closely related technique, electric field induced light scattering (EFLS) (13,14) has received less attention, but is also potentially useful for polymer characterization.

In the case of EFLS we observe the change in light scattering intensity when a strong electric field $\underset{\sim}{E}_o$ is applied across a polymer solution. The change occurs because the polymers are oriented and also deformed by the electric field. The deformation arises because the more polar conformations are energetically preferred over the less polar ones, in the presence of the external electric field. Consider a polymer solution of N polymer chains sufficiently dilute so that interference between the chains is negligible. Let each chain have n segments and total polarizability α. Then the change in VV scattering with the field on and off, $\Delta I_{vv}=I_{vv}(E)-I_{vv}(O)$ is (9),

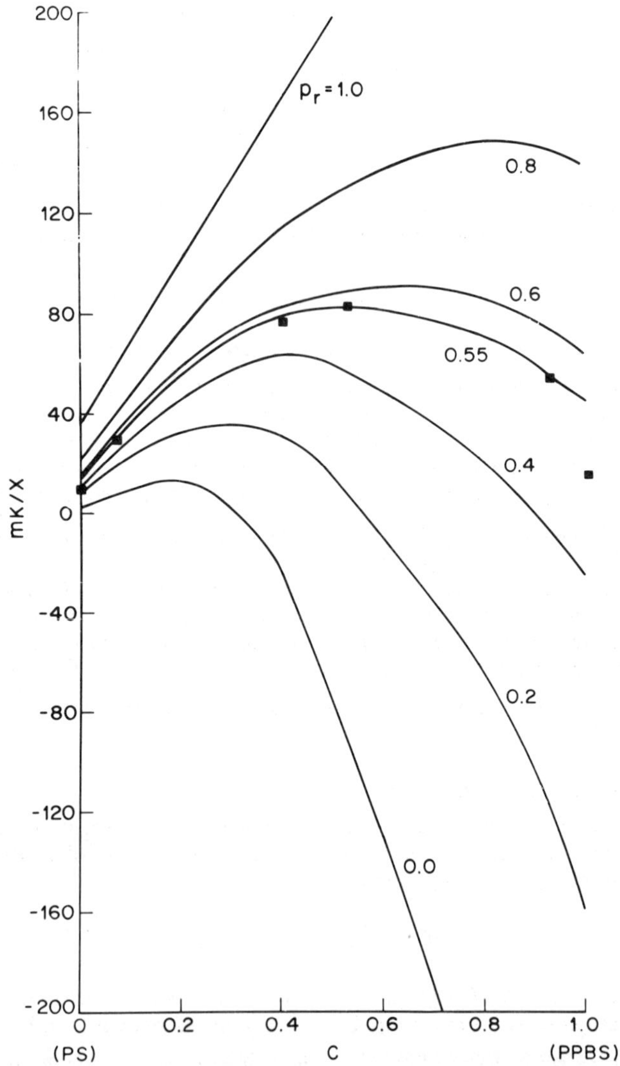

*Figure 1. Molar kerr constant values ($\times 10^{-12} cm^7 SC^{-2} mol^{-1}$) calculated for poly(styrene-co-*p*-bromostyrene) copolymers as a function of composition C and tacticity p_r. Black squares are experimental results. (Reproduced from Ref. 3. Copyright 1981, American Chemical Society.)*

$$\frac{\Delta I_{vv}}{E_o^2} = \frac{N\alpha^2}{2k^2T^2}\left\{\frac{1}{n^2}\sum_{i,j}\left[\frac{<j_o(qr_{ij})\mu^2>}{3} - <j_o(qr_{ij})>\frac{<\mu^2>}{3}\right.\right.$$

$$\left.\left. - <j_2(qr_{ij})\left(\frac{(\underset{\sim}{\mu}\cdot \underset{\sim}{r}_{ij})^2}{r_{ij}^2} - \frac{\mu^2}{3}\right)> P_2(\cos\Omega)\right]\right\} \quad (3)$$

We have assumed that the local and external field are equal. In eqn (3) $j_o(x)$ and $j_2(x)$ are the zeroth and second order spherical Bessel functions, $\underset{\sim}{r}_{ij}$ is the distance vector connecting the ith and jth bonds of the polymer, $P_2(x) = \frac{3x^2-1}{2}$ and Ω is the angle subtended between the scattering vector $\underset{\sim}{q}$ and the electric field $\underset{\sim}{E}_o$. The magnitude of q is $\frac{4\pi}{\lambda}\sin\frac{\theta}{2}$ and the summation extends over all elements i,j of the polymer. The angular brackets denote averaging over all internal configurations of the polymer. Eqn (3) shows that EFLS is a sensitive function of polymer flexibility and conformation. Few experimental results have been reported on flexible polymers. However Jennings (14) and coworkers have shown the sensitivity of EFLS to shape and flexibility for rigid and semi-rigid macromolecules.

Recently, Buckingham (15,16) and coworkers have predicted and observed a linear electrooptic effect, whereby an applied electric field induces a linear change in the intensity of light scattered by a gas in circularly polarized incident radiation. In this experiment one observes the difference in light scattering for left and right circularly polarized light in a plane normal to the applied electric field. This effect is a property of all molecules (chiral and achiral) and gives information about elements of the optical activity tensor $\underset{\sim}{G}$ and the dipole moment $\underset{\sim}{\mu}$. Experimental results have been reported for chloromethane (16) in the gas phase.

Dielectric Polarization and Saturation

The linear response of a medium to a weak applied electric field is characterized by the dielectric constant ϵ_o. From this experiment we deduce the molar polarization mP which is related to the mean square dipole moment $<\mu^2>$ by

$$mP = \frac{4\pi N_A}{3}\left(\overline{\alpha}^{\theta} + \frac{<\mu^2>}{3kT}\right) \quad (4)$$

$<\mu^2>$ has been used by us (1–5) and others to characterize the conformation of polymers, although it is not as sensitive to the details of polymer conformation as mK.

When a strong static electric field is applied across a medium, its dielectric and optical properties become anisotropic. When a low frequency analyzing electric field is used to probe the anisotropy, it is called the nonlinear dielectric effect (NLDE) or dielectric saturation (17). It is the low frequency analogue of the Kerr effect. The interactions which cause the NLDE are similar to those of EFLS. For a single flexible polar molecule, the external field will influence the molecule in two ways; firstly, it will interact with the total dipole moment and orient it, secondly, it will perturb the equilibrium conformation of the molecule to favor the conformations with the larger dipole moment. Thus, the orientation by the field will cause a decrease while the polarization of the molecule will cause an increase in the dielectric constant respectively, the net result depending on a sensitive interplay between the two effects.

Experimentally, one measures an incremental dielectric constant K. Thus

$$K = \epsilon_o + 3bE_o^2 + 5cE_o^4 + ... \tag{5}$$

where E_o is the static external field. A measure of the NLDE is the Piekara (18) factor,

$$\frac{\Delta\epsilon}{E_o^2} = \frac{K-\epsilon_o}{E_o^2} \tag{6}$$

From such an experiment we obtain the molar NLDE constant mS. Thus,

$$mS = 4\pi N_A \left[\frac{<tr(\hat{\tilde{\alpha}}^o\hat{\tilde{\alpha}}^o)>}{15kT} + \frac{2<\tilde{\mu}^T\hat{\tilde{\alpha}}^o\tilde{\mu}>}{15k^2T^2} + \frac{3<\mu^4> - 5<\mu^2>^2}{90k^3T^3} \right] \tag{7}$$

The first two terms arise from induced dipole and the last term from permanent dipole orientation, respectively. In figure 2 we present the experimental and calculated results for mS of the α,ω–dibromoalkanes, (16,19,20) and we find that the agreement is good.

It is interesting to note that the mS values calculated with the gauche states of all C-C bonds at ±120° (solid line) disagree with the NLDE constants measured for all the α,ω dibromoalkanes except 1,3-dibromopropane. When the terminal C-C bonds are permitted to adopt $\phi_{g\pm} = \pm 80°$, all other C-C bonds retaining gauche states at ±120°, the calculated NLDE constants (dashed lines) are in much better agreement with those observed for all the α,ω–dibromoalkanes except 1,3-dibromopropane. This behavior has a simple explanation in terms of the nonbonded steric and electrostatic interactions (21) resulting from rotation about the terminal C-C bonds.

When the terminal C-C bonds are in the gauche conformation, the Br atom and the γ carbon are in close proximity. In all α,ω–dibromoalkanes longer than 1,3-dibromopropane, the sterically bulky Br atom moves away from the γ carbon by adopting gauche states $<|120°|$ for the terminal C-C bonds. In 1,3-dibromopropane however, the γ carbon bears a partial charge of the same magnitude, but of opposite sign to that possessed by the Br atom. Thus, in the gauche states about the C-C bonds in 1,3-dibromopropane the attractive electrostatic interactions compensate the repulsive steric interactions between Br and C^γ, resulting in $\phi_{g\pm} = \pm 120°$. Clearly then, comparison of observed and calculated NLDE constants can provide detailed conformational information.

In Eqn (7) the last term is usually the most important for polar molecules and polymers. Therefore we may write to a good approximation that

$$mS = \frac{4\pi N_A <\mu^2>^2 \sigma}{30k^3T^3}$$

$$\sigma = \frac{<\mu^4>}{<\mu^2>^2} - \frac{5}{3} \tag{8}$$

σ is a measure of the deviation of a polymer from Gaussian statistics since $\sigma \rightarrow o$ as the number of repeat units x approaches infinity. σ differs from zero only when the chain is relatively short, say, fifty bonds or less. In table II we calculate (7) mS/x for the polymer poly(oxyethylene dimethyl ether) as a function of x.

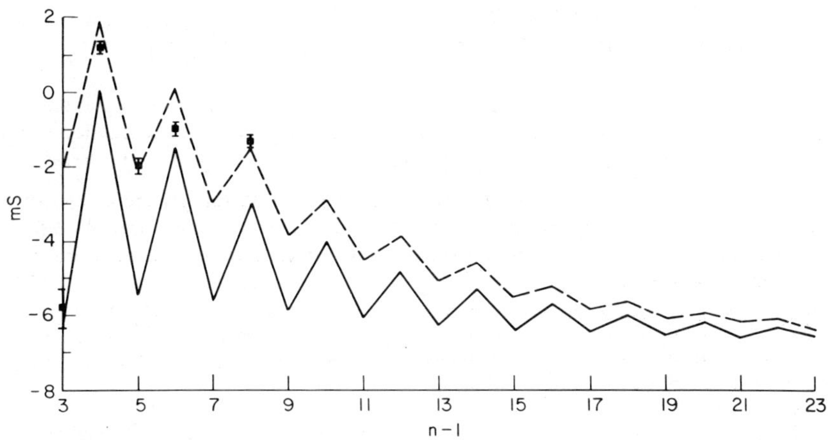

Figure 2. Nonlinear dielectric effect constant ($\times 10^{-8} cm^7 SC^{-2} mol^{-1}$) of $Br(CH_2)_{n-1}Br$ at 25 °C. Points represent the experimental results of Chelkowski (19, 20). *The solid and dashed lines are calculated with $\phi_{g\pm} = \pm 120°$ and $\pm 80°$, respectively, for the first and last C–C bonds. (Reproduced with permission from Ref. 6. Copyright 1981, American Institute of Physics.)*

Table II

mS/x(x10^{-9} cm^7 $SC^{-2}mol^{-1}$) and σ calculated for poly(oxyethylene dimethyl ether) at 25°C

$CH_3(OC_2H_4)_xOCH_3$

x	mS/x	10^3 σ
1	-16.4	-381
2	-7.70	-199
3	-2.87	-75
5	0.78	20
20	3.87	43
80	4.71	15

Calculations on other polymers and copolymers (7) show that the NLDE is sensitive to chain length, tacticity, copolymerization, and the details of the RIS model. Thus the NLDE is potentially useful to characterize the dielectric properties, microstructures and conformations of polymers.

Electric field Induced Second Harmonic Generation

When an intense laser beam of frequency ω passes through a medium lacking a center of symmetry, it is possible to observe second harmonic generation (SHG). In the case of an isotropic medium such as a liquid or a gas, the center of symmetry is removed by applying a strong electric field and the effect is called electric field induced second harmonic generation (EFSHG) (15,22). The effect depends on the square of the optical field strength E_ω and linearly on the static field E_0. The third order susceptibility of a single polymer chain to the optical and static fields is given by (22)

$$\gamma^a = \frac{1}{5}<\gamma_{\alpha\alpha\beta\beta}> + \frac{<\mu_\alpha\beta_{\alpha\beta\beta}>}{5kT} \qquad (9)$$

In eqn (9) $\gamma_{\alpha\alpha\beta\beta}$ transforms as a scalar. Clearly, when one uses the BAA, $\gamma_{\alpha\alpha\beta\beta}$ will be independent of conformation. We add the contribution of each bond in the molecule, as shown by Levine (22). γ can also be obtained from the optical Kerr effect. The second term in eqn (9) contains the product of two terms which transform as vectors. Thus we write

$$<\mu_\alpha\beta_{\alpha\beta\beta}> = <\underset{\sim}{\mu}\cdot\underset{\sim}{\beta}> \qquad (10)$$

The configurational average in eqn (10) can also be performed (8) with eqn (1). In table III we present the results of our calculations (8) of eqn (10) for poly(p-nitrostyrene) versus the tacticity of the polymer. No experimental EFSHG results have been reported for flexible polymers.

Table III

$\langle \underset{\sim}{\mu} \cdot \underset{\sim}{\beta} \rangle / x$ $(x10^{-48}\ cm^6)$ and $\langle \mu^2 \rangle / x$ $(x10^{-36} SC^2 cm^2)$

Calculated for poly(p-nitrostyrene) at 25 °C versus tacticity Pr; x = 200 repeat units

Pr	$\langle \underset{\sim}{\mu} \cdot \underset{\sim}{\beta} \rangle / x$	$\langle \mu^2 \rangle / x$
0.0 (isotactic)	21.5	10.7
0.2	16.6	8.27
0.4	16.3	8.12
0.6	16.8	8.36
0.8	16.4	8.17
1.0 (syndiotactic)	12.5	6.23

Conclusions

We have shown in this paper the relationships between the fundamental electrical parameters, such as the dipole moment, polarizability and hyperpolarizability, and the conformations of flexible polymers which are manifested in a number of their electrooptic and dielectric properties. These include the Kerr effect, dielectric polarization and saturation, electric field induced light scattering and second harmonic generation. Our experimental and theoretical studies of the Kerr effect show that it is very useful for the characterization of polymer microstructure. Our theoretical studies of the NLDE, EFLS and EFSHG also show that these effects are potentially useful, but there are very few experimental results reported in the literature with which to test the calculations. More experimental studies are needed to further our understanding of the nonlinear electrooptic and dielectric properties of flexible polymers.

Acknowledgment

One of us (G. K.) wishes to thank Professor R. S. Stein for his hospitality at the Polymer Research Institute, University of Massachusetts where this paper was written.

Literature Cited

1. Khanarian, G.; Tonelli, A. E. *J. Chem. Phys.*, 1981, 75, 5031.
2. Khanarian, G.; Tonelli, A. E. *Macromolecules*, 1982, 15, 145.
3. Khanarian, G.; Cais, R. E.; Kometani, J.; Tonelli, A. E.; *Macromolecules*, 1982, 15, 866.
4. Khanarian, G., Schilling, F. C.; Cais, R. E.; Tonelli, A. E.; *Macromolecules*, 1983, 16, 287.
5. Khanarian, G., Tonelli, A. E.; Cais, R. E., in preparation.
6. Khanarian, G. *J. Chem. Phys.* 1982, 76, 3186.
7. Khanarian, G. *Macromolecules* 1982, 15, 1429.

8. Khanarian, G. *J. Chem. Phys.* 1982, 77, 2684.

9. Khanarian, G.; Stein, R. S. in preparation.

10. Flory, P. J. 'Statistical Mechanics of Chain Molecules'. Interscience: New York, 1969.

11. Miller, C. K.; Orr, B. J.; Ward, J. F. *J. Chem. Phys.* 1981, 74, 4858.

12. Buckingham, A. D.; Pople, J. A. *Proc. Phys. Soc. (London)* 1955, A68, 905.

13. Wippler, C.; Benoit, H. *Makromol. Chem.* 1954, 13, 7.

14. Jennings, B. R. *Pure & Appl. Chem.* 1982, 54, 395.

15. Buckingham, A. D. *J. Phys. Chem.* 1982, 86, 1175.

16. Buckingham, A. D.; Shatwell, R. A. *Phys. Rev. Lett.* 1980, 45, 21.

17. Buckingham, A. D. *J. Chem. Phys.* 1956, 25, 428.

18. Piekara, A. *Proc. R. Soc. London. Ser. A.* 1939, 172, 360.

19. Chelkowski, A. Dielectric Physics (Polish Scietific, Warsaw 1980) Chap. 4, p. 240.

20. Chelkowski, A. Fiz. Dielektr. (Poznan) 1962, 1,3.

21. Tonelli, A. E. Macromolecules, 1982, 15, 290.

22. Levine, B. F. in 'Dielectric and Related Molecular Processes,' Davis, M., ed. Chemical Society (London), 1977, 3, 73.

RECEIVED May 16, 1983

INDEX

INDEX

Production by Deborah Corson and Anne Riesberg
Indexing by Deborah Corson
Jacket design by Anne Bigler

Elements typeset by Hot Type Ltd., Washington, DC
Printed and bound by Maple Press Co., York, PA